ASSOCIATIVE FUNCTIONS

Triangular Norms and Copulas

ASSOCIATIVE FUNCTIONS

Triangular Norms and Copulas

Claudi Alsina
Universitat Politècnica de Catalunya, Barcelona, Spain

Maurice J Frank
Illinois Institute of Technology, Chicago, USA

Berthold Schweizer
University of Massachusetts, Amherst, USA

NEW JERSEY • LONDON • SINGAPORE • BEIJING • SHANGHAI • HONG KONG • TAIPEI • CHENNAI

Published by

World Scientific Publishing Co. Pte. Ltd.
5 Toh Tuck Link, Singapore 596224
USA office: 27 Warren Street, Suite 401-402, Hackensack, NJ 07601
UK office: 57 Shelton Street, Covent Garden, London WC2H 9HE

British Library Cataloguing-in-Publication Data
A catalogue record for this book is available from the British Library.

ASSOCIATIVE FUNCTIONS: TRIANGULAR NORMS AND COPULAS

ISBN-13 978-981-256-671-3
ISBN-10 981-256-671-6

Printed in Singapore

TO

CARME & (PAT & JUDIE)

OR

(CARME & PAT) & JUDIE

Preface

Associativity is an ancient concept. However, the modern theory of the functional equation of associativity on the real line begins with a celebrated paper of J. Aczél, written in 1949, in which he gives a general representation theorem for associative functions on intervals.

The impetus for this book stems from the theory of probabilistic metric spaces. There, in connection with the triangle inequality, it became necessary to have large classes of associative functions at one's disposal, to know the properties of individual associative functions, the relationship between pairs of associative functions, etc. Subsequently, the same need arose in the Kampé de Fériet-Forte theory of information without probability, in theories of multivalued logic and fuzzy sets and in various problems in statistics.

By the early 1980's, problems centered around the associativity equation were receiving considerable attention, the literature on the subject was growing but widely scattered, and many facts that were well known to experts in the field were constantly being rediscovered by newcomers. Accordingly, at an International Symposium on Functional Equations which was held in Oberwolfach in December 1984, we decided to gather the basics of the theory in one place and, in the process, simplify many proofs and add a number of new results. But then, for personal and professional reasons (e.g., the first two authors became involved with heavy academic responsibilities and, more recently, the third author was occupied with the editing of the "Karl Menger Selecta Mathematica"), the work bogged down, at times to a standstill. In the meantime, an explosion of interest in and work on associative functions (in particular, t-norms and associative copulas) was taking place. Numerous papers appeared; a special issue of Fuzzy Sets and Systems (Vol. 104, No. 1, May 1999) was devoted to the topic;

the essentials were presented in chapters of several books, e.g., [Gottwald (1993); Fodor and Roubens (1994b); Nguyen and Walker (2000); Nelsen (1999); Hadzic and Pap (2001)]. These, as their titles indicate, are more concerned with applications than with fundamentals. Then, in 2000, the book "Triangular Norms" by E.P. Klement, R. Mesiar and E. Pap was published. Surprisingly – and contrary to what one might expect – aside from the very basic facts, many of which are contained in the book "Probabilistic Metric Spaces" by the third author and A. Sklar, there is little overlap between our book and the book by Klement, Mesiar and Pap. Indeed, the two books complement each other very well. Thus, in spite of the passage of time, we can still say that the contents of this book – which have of course been brought up to date and which include many results not heretofore published – are the foundation on which all the other developments rest.

This book is divided into four chapters. Chapter 1 is introductory. In it, we first present a brief overview of the basic facts concerning some of the classical functional equations and the associated inequalities: they all play an important role in our studies. The bulk of the chapter is devoted to introducing the associative functions which will be our primary concern in the remainder of the book, the so-called t-norms, and two other classes of functions, s-norms and copulas. We define these functions, derive some of their basic properties, and establish some of the relations within and among these classes. In this chapter, we also establish the terminology and the notational conventions which we use in the sequel.

Chapter 2 is devoted to the basic representation theorem for associative functions and some of its consequences. We prove this theorem in the form which is most suitable for our purposes and discuss various other versions as well as generalizations. This chapter also includes a table in which we list a number of one-parameter families of frequently encountered t-norms, together with their properties. Chapters 3 and 4 are devoted to functional equations and inequalities that involve associative functions. Many of the results presented here stem from problems that arose in connection with further developments of the theory.

The book concludes with two appendices and an extensive bibliography. In Appendix A, we list examples and counterexamples that illuminate the text. In Appendix B, we present a series of open problems for further research.

This book is intended to be primarily a reference work. Nevertheless, it is also suitable for use as a text for a one-semester advanced undergraduate or beginning graduate course on functional equations. True, we deal primarily with one class of functions and a restricted class of functional equations. However, in the process, we employ most of the standard techniques – and some new ones – for solving functional equations and many of the basic equations and inequalities are encountered in the process. Thus, it can be argued that the fact that this book is focused on one central theme is an advantage rather than a hindrance.

We are grateful to Prof. A. Sklar (Chicago, Illinois), Prof. J. Aczél (Waterloo, Ontario) and Prof. W. Sander (Clausthal, Germany) for their critical remarks, to Prof. A. Monreal (Barcelona, Spain) for making the illustrations included in this book and to Mrs. Rosa Navarro (Barcelona, Spain) for her efficient typing, of various versions, of our manuscript.

Special Symbols

$\circ$	composition of functions
C	copula
$C^{\wedge}$	dual copula
A^c	complement of the set A
$df(X)$	distribution function of the random variable X
Dom	domain
δ_T	diagonal function of a t-norm T
Δ	set of distribution functions
Δ^+	elements F in Δ with $F(0) = 0$
f^{-1}	inverse function of f
f^n	n-th iterate of f
$\mathcal{F}[a,b]$	set of associative functions on $[a,b]$
$\mathcal{G}[a,b]$	set of associative functions on $[a,b]$
I	closed unit interval [0,1]
I^n	n-th cartesian product of I
J	generic interval
j_A	identity function on the set A
Max	Maximum
Min	Minimum
P	product $P(x,y) = xy$, for x, y in I
$\mathbb{R}$	extended real line $[-\infty, \infty]$
$\mathbb{R}^+$	extended positive half-line $[0, \infty]$
Ran	range

S	s-norm
S^*	t-norm associated with an s-norm S
$\mathcal{S}$	set of s-norms
$\mathcal{S}_{Ar}$	set of continuous Archimedean s-norms
$\mathcal{S}_{Co}$	set of continuous s-norms
$\mathcal{S}_{St}$	set of strict s-norms
σ_x	vertical section
σ_y	horizontal section
T	t-norm
T^*	s-norm associated with a t-norm T
$\mathcal{T}$	set of t-norms
$\mathcal{T}_{Ar}$	set of continuous Archimedean t-norms
$\mathcal{T}_{Co}$	set of continuous t-norms
$\mathcal{T}_{St}$	set of strict t-norms
W	$W(x, y) = \max(x + y - 1, 0)$, for x, y in I
χ_A	indicator function of the set A
Z	minimal t-norm
$\mathcal{Z}(T)$	zero set of a t-norm T

Contents

Preface vii

Special Symbols xi

1. Introduction 1

1.1 Historical notes . . . 1
1.2 Preliminaries . . . 6
1.3 t-norms and s-norms . . . 9
1.4 Copulas . . . 17

2. Representation theorems for associative functions 23

2.1 Continuous, Archimedean t-norms . . . 23
2.2 Additive and multiplicative generators . . . 38
2.3 Extension to arbitrary closed intervals . . . 51
2.4 Continuous, non-Archimedean t-norms . . . 57
2.5 Non-continuous t-norms . . . 64
2.6 Families of t-norms . . . 70
2.7 Other representation theorems . . . 81
2.8 Related functional equations . . . 93

3. Functional equations involving t-norms 99

3.1 Simultaneous associativity . . . 99
3.2 n-duality . . . 110
3.3 Simple characterizations of Min . . . 127
3.4 Homogeneity . . . 129
3.5 Distributivity . . . 134

3.6 Conical t-norms 137
3.7 Rational Archimedean t-norms 143
3.8 Extension and sets of uniqueness 151

4. Inequalities involving t-norms 173

4.1 Notions of concavity and convexity 173
4.2 The dominance relation 182
4.3 Uniformly close associative functions 189
4.4 Serial iterates and n-copulas 194
4.5 Positivity 203

Appendix A Examples and counterexamples 209

Appendix B Open problems 219

Bibliography 223

Index 235

Chapter 1

Introduction

1.1 Historical notes

The story of the functional equation of associativity begins with Abel. The first paper that he published in Crelle's Journal (volume 1, 1826, pp. 11-15) is entitled "Untersuchung der Function zweier unabhängig veränderlichen Grössen x und y wie $f(x,y)$, welche die Eigenschaft haben, dass $f(z,f(x,y))$ eine symmetrische Function von x, y und z ist". In this paper, which has been called "the earliest semigroup paper" [Lawson (1996)], Abel showed that if f is differentiable and satisfies the system of functional equations

$$\begin{aligned} f(x,f(y,z)) &= f(z,f(x,y)) = f(y,f(z,x)) = f(x,f(z,y)) \\ &= f(z,f(y,x)) = f(y,f(x,z)) \end{aligned}$$

then there exists a differentiable and invertible function ψ such that

$$\psi(f(x,y)) = \psi(x) + \psi(y).$$

Scant attention was given to Abel's paper in the rest of the 19th century. Then, in 1900, Hilbert supplied a strong impetus via the 5th problem of the famous address he delivered at the International Congress of Mathematicians in Paris. (See the English translation which is reprinted in Volume 28 of the Proc. of Symposia in Pure Math., Part I, pp. 1-34.) This problem consists of two parts. In the first part, Hilbert poses the question: "How far Lie's concept of continuous groups of transformations is approachable in our investigations without the assumption of differentiability of the functions". In the second, referring specifically to Abel, he asks: "In how far are the assertions which we can make in the case of differentiable functions true under proper modifications without this assumption?".

Of the two parts of Hilbert's problem, the first is by far the better known and has received much more attention. For the one-dimensional case, solutions were given by L.E.J. Brouwer [1909] and then by E. Cartan [1930]; and the problem as posed was completely solved by A. Gleason, D. Montgomery and L. Zippin in 1952. Work on its ramifications continues to this day. An attack on the second part of Hilbert's problem had to wait for the maturation of the theory of functional equations.

During the first half of the 20th century much of the work on functional equations was sporadic and scattered throughout the mathematical literature. A large number of papers were concerned with the equations that are associated with the names of Cauchy and Jensen; and here and there one finds a paper dealing with the associativity equation

$$T(x, T(y, z)) = T(T(x, y), z). \tag{1.1.1}$$

(For details, see the classic treatise [Aczél (1966)] and the various bibliographies on "Works on Functional Equations" in the journal Aequationes Mathematicae.)

The modern theory of the associativity equation (1.1.1) – indeed much of the modern theory of functional equations involving multiplace functions, together with a serious attack on the second part of Hilbert's fifth problem (see [Aczél (1989)]) – begins with the pioneering paper of J. Aczél [1949]. In it he proves the following

Theorem *Let J denote any proper subinterval of the extended real line $\overline{\mathbb{R}}$. Then the function $A : J \times J \to J$ is associative, continuous and cancellative if and only if it admits the representation*

$$A(x, y) = a^{-1}(a(x) + a(y)), \tag{1.1.2}$$

where $a : J \to \overline{\mathbb{R}}$ is continuous and strictly monotonic. Moreover, J must be open or half-open.

Aczél's Theorem improved earlier versions of the representation (1.1.2) by eliminating hypotheses such as commutativity, the presence of an identity, an underlying group structure and, most emphatically, differentiability. It and related results have provided much of the stimulus for the growth of the field of functional equations in the last half-century (see, e.g., the books [Aczél (1966); Kuzcma (1968); Dhombres (1979); Targonski (1981); Kuczma (1985); Aczél (1987); Smítal (1988); Aczél and Dhombres (1989);

Kuczma, Choczewski and Ger (1990)], Chapters 5-7 of [Schweizer and Sklar (1983), (2005)], and the journals Publicationes Mathematicae Debrecen and Aequationes Mathematicae).

While Aczél and his co-workers were developing the theory of functional equations in the spirit of Abel and Cauchy, another group of mathematicians, under the guidance of A.D. Wallace and A.H. Clifford, was extending the work of Brouwer, Cartan et al. from topological groups to topological semigroups (see, e.g., [Faucett (1955); Mostert and Shields (1957); Clifford (1958); Clifford and Preston (1961), (1967); Fuchs (1963); Hofmann and Mostert (1966); Paalman-de Miranda (1970); Carruth, Hildebrant and Koch (1983)], as well as the recent and interesting surveys [Hofmann (1994); Lawson (1996); Hofmann and Lawson (1996)]).

In 1965, the two roads met when C.-H. Ling [1965], motivated by Aczél's theorem and starting from somewhat different hypotheses, extended the representation (1.1.1) from strictly increasing functions to non-decreasing Archimedean functions (with identity) and then, via the so-called ordinal sums, to all non-decreasing functions. For some extensions of Ling's results see [Krause (1981)] and Section 2.7.

The abovementioned results close one chapter of the study of continuous associative functions on intervals and in a linearly ordered set. The results of Mostert and Shields characterize such functions up to order-isomorphism; and the results of Aczél and Ling yield a universal representation of all such functions. But to say that the story ends here would be analogous to saying that the theory of second order linear differential equations with constant coefficients ends with the general solution of $ay''+by'+c=0$. For, together with the need for the general solution, there is also a need for particular solutions, particular families of solutions, and solutions having particular properties. Such a need was already evident in the work of T. S. Motzkin [1936], who was interested in finding those associative functions that can be used to define metrics on Cartesian products of metric spaces, and A. Bohnenblust [1940], who was interested in finding the associative functions that characterize norms on certain linear spaces. This need also arose some years later when J. Kampé de Fériet and B. Forte [1967] laid the foundations of the theory of information without probability: here associative functions on the positive half-line, the so-called composition laws, play a crucial role. It is also evident in the studies that eventually led us to write this book.

The origins of this book can be traced back to two papers by the third author and A. Sklar. In the first paper [Schweizer and Sklar (1960)], they laid the foundations of the theory of probabilistic metric spaces, first defined by K. Menger [1942]. These are generalizations of metric spaces in which the distances between points are described by probability distributions rather than by numbers. (See the book [Schweizer and Sklar (1983), (2005)] as well as the recent survey [Schweizer (2003)]). With each pair of points, p and q, in such a space, there is associated a probability distribution function F_{pq} whose value $F_{pq}(x)$, for any real number x, is usually interpreted as the probability that the "distance" between p and q is less than x. The triangle inequality in such spaces takes the form

$$F_{pr} \geq \tau(F_{pq}, F_{qr}), \tag{1.1.3}$$

where τ is a suitable continuous semigroup operation on the space of distribution functions. The most commonly occurring "triangle functions" τ have the form

$$\tau_T(F, G)(x) = \sup_{u+v=x} T(F(u), G(v)), \tag{1.1.4}$$

where T is a continuous "t-norm", i.e., a suitable continuous semigroup operation on the unit interval [0,1]. Different triangle functions generally lead to probabilistic metric spaces with different geometric and topological properties; and different t-norms lead to different triangle functions. Thus, in order to penetrate deeply into the structure theory of probabilistic metric spaces, it is necessary to have a repertoire of triangle functions and t-norms at hand. This need was already apparent in the abovementioned first paper and was clearly expressed in the second [Schweizer and Sklar (1961)]. In that paper, Aczél's Theorem was used to construct several one-parameter families of t-norms, to derive various properties and relations among t-norms, and to analyze the triangle inequality (1.1.3). Last but not least, it was used to relate t-norms to another family of two-place functions, namely the copulas which had been defined earlier by A. Sklar [1959] (see [Sklar (1996a)]). These are functions that link (two-dimensional) probability distribution functions to their one-dimensional margins. They play an important role in the theory of probabilistic metric spaces and, in recent years, have turned out to be increasingly significant in statistics, particularly in the study of concepts and measures of dependence (see [Schweizer (1991)] and the monograph [Nelsen (1999)]).

In 1965, L. Zadeh founded the theory of fuzzy sets (although the original definition goes back to Menger [1951b]) and used the associative functions Maximum and Minimum to define generalized unions and intersections. Subsequently, this work was related to the multivalued logic of

J. Lukasiewicz; and later, in generalizations of the Lukasiewicz theory, as well as in the theory of fuzzy sets itself, Minimum and Maximum were replaced by a t-norm and its corresponding s-norm (t-conorm), respectively [Alsina, Trillas and Valverde (1980), (1983); Dubois (1980); Höhle, private communication; Klement (1980), (1982)]. This work has led to a number of interesting functional equations and inequalities, many of which have been studied by the first author, E. Trillas and their colleagues (see, e.g., [Alsina (1985b), (1988), (1996); Trillas (1980); Trillas and Valverde (1982), (1984); Alsina, Trillas and Valverde (1982)]). The results obtained have found application in the fields of artificial intelligence and cluster analysis [Ruspini (1982); López de Mantaras and Valverde (1984); Zimmermann (1991); Höhle and Klement (1995)] as well as in the theory of synthesis of judgements [Aczél and Saaty (1983); Aczél and Alsina (1984), (1986), (1987); Aczél (1987)]. In addition, t-norms, s-norms and other associative functions have come to play a pervasive role in the theory of fuzzy sets, multivalued logics and their ramifications. Listing and discussing all of these matters here would take us too far afield. Fortunately, we do not have to: they are reviewed and developed in detail in the book [Klement, Mesiar and Pap (2000)].

In the last years of the 20th century and in the first years of this century research on t-norms and related functions has continued at a steady pace. Here we mention the papers [Klement, Mesiar and Pap (2004a), (2004b), (2004c)] in which the salient facts about t-norms presented in the book [Klement, Mesiar and Pap (2000)], together with some more recent results, are summarized; the lists of open problems [Alsina, Frank and Schweizer (2003); Klement, Mesiar and Pap (2004d)] and lastly, the conference proceedings [Klement and Mesiar (2005)] which give an overview of t-norms and their applications.

The theory of probabilistic metric spaces has continued to be a source of interesting problems. As indicated previously, in that theory the triangle inequality takes the form (1.1.3), and most triangle functions are induced by semigroup operations on [0,1], e.g., as in (1.1.4). In the other direction, many questions concerning such triangle functions can be reduced to questions concerning associative functions on [0,1]. For example, when F and G are step-functions, any functional equation (such as the autodistributivity equation) involving the function τ_T defined by (1.1.4) reduces to a functional equation, or a system of functional equations, for T. Questions of this kind have motivated much of the work of the first author [Alsina (1980), (1981), (1984a)]. Also, in [Frank (1975), (1979), (1991)], the second author studied a family of binary operations on the space of probability distribution functions which are related to sums of dependent random variables and are induced by copulas. This led him to the functional equation of simultaneous associativity and to the discovery of a one-parameter family of associative copulas which, in recent years, have turned out to be of great importance in certain areas of statistics [Genest and MacKay (1986a), (1986b); Genest (1987); Nelsen (1986), (1991), (1995), (1997)]. Additional examples of this interplay may be found in the text.

One final comment: In the early papers on probabilistic metric spaces, the triangle inequality (1.1.3) had the form $F_{pr}(x+y) \geq T\left(F_{pq}(x), F_{qr}(y)\right)$ and, following Menger [1942], the function T was called a triangular norm, or briefly, a t-norm. Today t-norms arise in many situations where there are no triangles to be found and the name has become an anachronism – and a somewhat misleading one at that. However, this usage of the term is now prevalent in much of the literature, and so, after much agonizing (and after a number of futile attempts to find a reasonable alternative), we have decided not to tamper with it. On the other hand, we have decided to replace the appellation "t-conorm" by "s-norm".

1.2 Preliminaries

In this section we present the notational and other conventions used in this book, as well as some basic prerequisites from the theory of functional equations and inequalities.

Generally, the letters u, v, w will denote elements of the extended real line $\mathbb{R} := [-\infty, \infty]$; x, y, z will denote real numbers in the closed unit interval $I := [0, 1]$, while a, b, c, d will be used to denote endpoints of real intervals. In particular, the set of non-negative reals is $\mathbb{R}^+ := [0, \infty]$. Real one-place functions will be represented by lower case letters, e.g., f, g, h, t, s; the symbol $\circ$ will denote composition and j_A will denote the identity function on the set A (the reference to A will be supressed whenever there is no chance of confusion). When Dom $f \subseteq$ Ran f, f^n will denote the n-th iterate of f, defined recursively by

$$f^0 = j_{\mathrm{Dom}\ f} \quad \text{and} \quad f^n = f \circ f^{n-1} \quad \text{for } n \geq 1;$$

and, when it exists, the inverse of f will be denoted by f^{-1}.

Binary operations on $\mathbb{R}$ will be denoted by capital letters, e.g., F, G, H, S, T, and will be treated as two-place functions, i.e., we adopt the functional approach rather than the algebraic one. Also, functions will be allowed to take on the values $\pm\infty$ at boundary points of their domains.

The most fundamental functional equation is *Cauchy's functional equation*,

$$f(x+y) = f(x) + f(y). \tag{1.2.1}$$

For functions $f : \mathbb{R} \to \mathbb{R}$ which satisfy mild regularity conditions (continuity at a single point, or monotonicity, or boundedness from above or below on a set of positive measure, etc.) the general solution of (1.2.1) is

$$f(x) = cx, \tag{1.2.2}$$

where c is an arbitrary constant. If no regularity condition is assumed, then (1.2.1) admits solutions that are discontinuous everywhere and whose graphs are dense in the plane; indeed, the general solution can be expressed in terms of the values of f on a Hamel basis of $\mathbb{R}$. Moreover, there may be other solutions of Cauchy's equation when (1.2.1) is assumed to hold conditionally, i.e., only for points (x, y) in a subset of $\mathbb{R}^2$ [Kuczma (1978), (1985)].

Three other equations of Cauchy type,

$$\begin{aligned} f(x+y) &= f(x)f(y), \\ f(xy) &= f(x) + f(y), \\ f(xy) &= f(x)f(y), \end{aligned} \tag{1.2.3}$$

can be reduced to (1.2.1) by appropriate changes of variables. Under any of the regularity conditions mentioned above, their non-trivial solutions are given by $f(x) = e^{cx}$, $c \log x$, and x^c, on $\mathbb{R}$, $\mathbb{R}^+$, and $\mathbb{R}^+$, respectively.

Another important functional equation is *Jensen's functional equation*,

$$f\left(\frac{x+y}{2}\right) = \frac{f(x) + f(y)}{2}, \tag{1.2.4}$$

whose general solution, under weak regularity conditions, is

$$f(x) = ax + b,$$

where a and b are arbitrary constants.

Equations (1.2.1) and (1.2.4) are closely related to certain basic inequalities. Equation (1.2.1) can be viewed as a special case of either

$$f(x+y) \leq f(x) + f(y) \tag{1.2.5}$$

or

$$f(x+y) \geq f(x) + f(y). \tag{1.2.6}$$

When (1.2.5) or (1.2.6) holds, f is said to be *subadditive* or *superadditive*, respectively. Equation (1.2.4) leads to the study of

$$f\left(\frac{x+y}{2}\right) \leq \frac{f(x)+f(y)}{2} \tag{1.2.7}$$

and

$$f\left(\frac{x+y}{2}\right) \geq \frac{f(x)+f(y)}{2}. \tag{1.2.8}$$

When (1.2.7) or (1.2.8) holds, f it is said to be *midpoint* (or *Jensen*) *convex* or *concave*, respectively. Under certain regularity conditions these midpoint properties extend to full convexity and concavity, respectively, i.e., for all λ in I,

$$f(\lambda x + (1-\lambda)y) \leq \lambda f(x) + (1-\lambda) f(y) \tag{1.2.9}$$

or

$$f(\lambda x + (1-\lambda)y) \geq \lambda f(x) + (1-\lambda) f(y). \tag{1.2.10}$$

There are connections between the above inequalities; for instance, if f is concave and $f(0) = 0$, then f is subadditive. Most of the important inequalities of classical real analysis are based on one or another of these inequalities.

A functional equation in a single variable which plays a key role in iteration theory is *Schröder's functional equation*,

$$f(g(x)) = sf(x), \tag{1.2.11}$$

where g is a given function and s is a constant. Under appropriate conditions on g, the general solution of (1.2.11) depends on an arbitrary function defined on a subinterval of Dom f which can be extended to Dom f by way of the iterated equation

$$f(g^n(x)) = s^n f(x).$$

A similar situation prevails in the case of *Abel's functional equation*,

$$f(g(x)) = a + f(x). \tag{1.2.12}$$

For detailed studies of these equations, consult [Aczél (1966), (1987); Aczél and Dhombres (1989); Kuczma (1968), (1985); Dhombres (1979); Kuczma, Choczewski and Ger (1990)]; for the inequalities, see [Hardy, Littlewood and Polya (1952); Beckenbach and Bellman (1965); Marshall and Olkin (1979)].

1.3 t-norms and s-norms

The principal objects of our study are the associative binary operations (semigroup operations) on $I = [0,1]$ that are order-preserving and commutative and have identity 1 or 0. Throughout, we generally adopt function notation and terminology for algebraic systems. Thus, for instance, a semigroup operation T on I is viewed as a two-place function from I^2 to I that satisfies the functional equation of associativity (1.1.1). This section is devoted to presenting the most basic definitions, terminology, notation, and examples, and to establishing a number of elementary results.

Definition 1.3.1 A **t-norm** is a two-place function $T : I^2 \to I$ (i.e., a binary operation on I) which satisfies the following conditions:

(i) On the boundary of I^2,

$$T(x,0) = T(0,x) = 0, \tag{1.3.1a}$$

$$T(x,1) = T(1,x) = x. \tag{1.3.1b}$$

(ii) T is non-decreasing in each place, i.e.,

$$T(x_1,y_1) \le T(x_2,y_2), \text{ whenever } x_1 \le x_2, y_1 \le y_2. \tag{1.3.2}$$

(iii) T is commutative, i.e., for all x, y in I,

$$T(x,y) = T(y,x). \tag{1.3.3}$$

(iv) T is associative, i.e., for all x, y, z in I,

$$T\left(T(x,y),z\right) = T(x,T(y,z)). \tag{1.3.4}$$

Algebraically, a t-norm is a commutative, order-preserving semigroup operation on [0,1] with identity 1 and null element 0. Geometrically, the graph of a t-norm is a surface over the unit square which is bounded by the quadrilateral whose vertices are $(0,0,0)$, $(1,0,0)$, $(1,1,1)$ and $(0,1,0)$, which rises both horizontally and vertically, and which is symmetric with respect to the plane $x = y$. (See Figure 1.3.1.) Associativity seems to have no simple geometric interpretation.

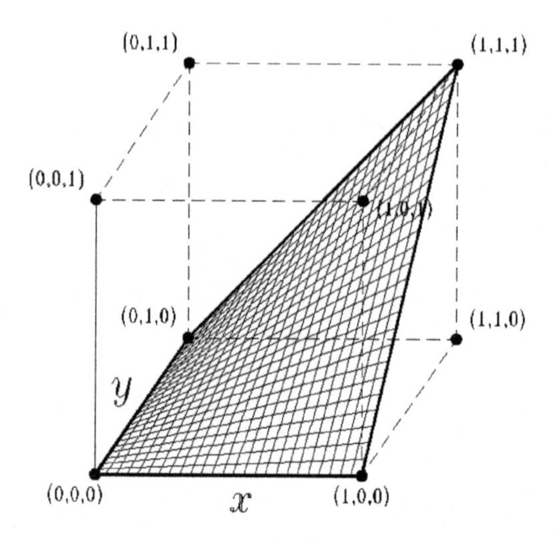

Figure 1.3.1. The graph of a t-norm.

In view of the monotonicity and commutativity of T and the fact that Ran $T \subseteq I$, the boundary conditions (1.3.1a) and (1.3.1b) can be replaced by the single condition:

(v) For all x in I,

$$T(x, 1) = x.$$

For then $0 \leq T(x, 0) \leq T(1, 0) = 0$, so that $T(x, 0) = 0$, and the remaining conditions follow from (1.3.3). On the other hand, Examples A.1.1, A.1.2, A.1.3, and A.1.4 of Appendix A show that the conditions (1.3.1), (1.3.2), (1.3.3), and (1.3.4) are independent.

In the presence of continuity the situation changes dramatically. Thus if T is continuous and satisfies (1.3.1) and (1.3.4), then T is non-decreasing and commutative, i.e., satisfies (1.3.2) and (1.3.3). (See Lemma 2.1.1 and Corollary 2.1.7.).

Definition 1.3.2 A t-norm T is **strict** if it is continuous on I^2 and strictly increasing in each place on $(0,1]^2$, so that

$$T(x_1,y) < T(x_2,y), \quad \text{whenever } x_1 < x_2,\ y > 0,$$

and $$T(x,y_1) < T(x,y_2), \quad \text{whenever } x > 0,\ y_1 < y_2. \tag{1.3.5}$$

In the sequel we let $\mathcal{T}$ denote the set of t-norms, $\mathcal{T}_{Co}$ the set of continuous t-norms, and $\mathcal{T}_{St}$ the set of strict t-norms. Thus $\mathcal{T}_{Co}$ consists of the t-norms T for which (I,T) is a topological semigroup and, as is readily seen, $\mathcal{T}_{St}$ those for which the cancellation law also holds. In the literature on topological semigroups, the elements of $\mathcal{T}_{Co}$ are known as I-semigroups [Paalman-de Miranda (1964); Carruth, et al. (1983)].

The natural ordering on I induces a partial ordering on the set of all functions from I^2 to I and, in particular, on the set of all t-norms. Accordingly, we say that T_1 is **weaker** than T_2, or T_2 is **stronger** than T_1, and we write $T_1 < T_2$, if

$$T_1(x,y) \le T_2(x,y), \qquad \text{for all } (x,y) \text{ in } I^2,$$

and

$$T_1(x_0,y_0) < T_2(x_0,y_0), \qquad \text{for some } (x_0,y_0) \text{ in } I^2.$$

If $T_1 < T_2$ or $T_1 = T_2$, we write $T_1 \le T_2$.

The most important t-norms, which we designate by the symbols Z, W, P, and Min, are given by

$$\begin{aligned} Z(x,y) &= \begin{cases} x, & y = 1, \\ y, & x = 1, \\ 0, & \text{otherwise}, \end{cases} \\ W(x,y) &= \text{maximum } (x+y-1, 0), \\ P(x,y) &= xy, \\ \text{Min}(x,y) &= \text{minimum } (x,y). \end{aligned} \tag{1.3.6}$$

The graphs of these t-norms are depicted in Figure 1.3.2. Note that W, P, and Min are in $\mathcal{T}_{Co}$, that only P is in $\mathcal{T}_{St}$, and that

$$Z < W < P < \text{Min}. \tag{1.3.7}$$

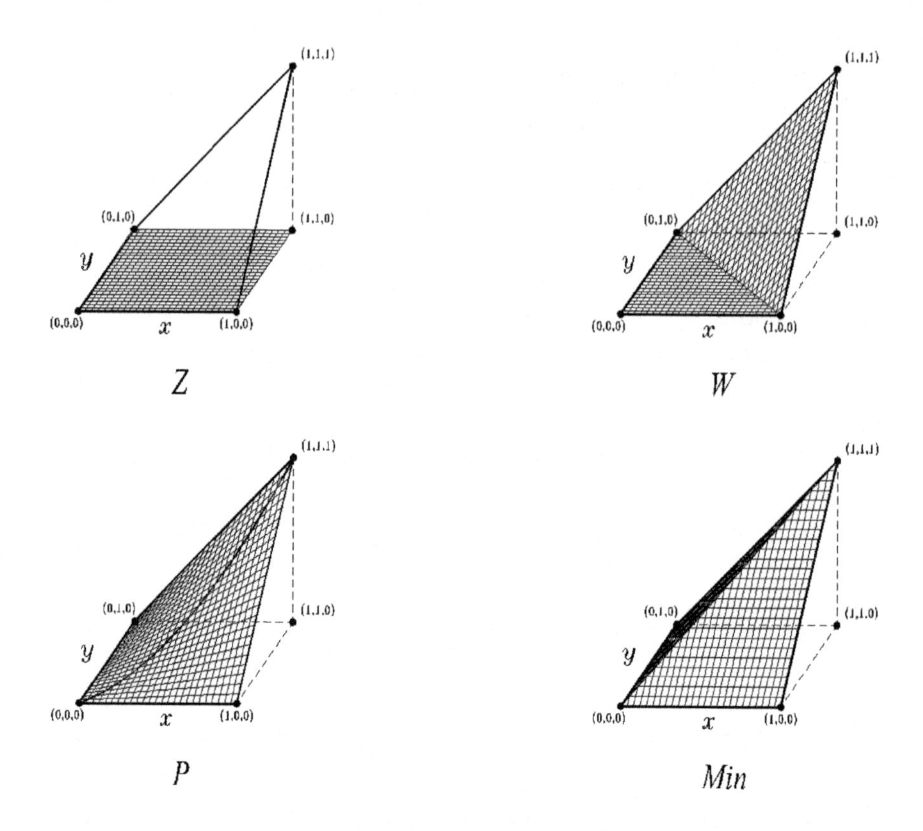

Figure 1.3.2. Graphs of the t-norms Z, W, P, Min.

Lemma 1.3.3 *If $T : I^2 \to I$ satisfies (1.3.1) and (1.3.2), and in particular if T is a t-norm, then*

$$Z \leq T \leq Min. \tag{1.3.8}$$

These inequalities follow at once from the inequalities

$$0 \leq T(x, y) \leq T(x, 1) \text{ and } 0 \leq T(x, y) \leq T(1, y),$$

and (1.3.1b). Note, in particular, that Z is the weakest and Min the strongest t-norm, or in other words that Z and Min are, respectively, the infimum and the supremum of the poset $(\mathcal{T}, \leq)$. But since neither the infimum nor the supremum of two associative functions need be associative (see Examples A.9.2, A.9.3 of Appendix A), $(\mathcal{T}, \leq)$ is not a lattice, nor even a semilattice.

Since T is associative, for each x in I, the **T-powers** of x may be defined recursively by

$$x^1 = x \text{ and } x^{n+1} = T(x^n, x), \tag{1.3.9}$$

for all positive integers n. Straightforward inductions yield that, for all $m, n \geq 1$,

$$x^{m+n} = T(x^m, x^n) = T(x^n, x^m), \tag{1.3.10}$$

and

$$x^{mn} = (x^m)^n = (x^n)^m. \tag{1.3.11}$$

Moreover, (1.3.9) can be extended to non-negative integers by defining $x^0 = 1$ for all $x \neq 0$. When this is done, then for all $x \neq 0$, (1.3.10) and (1.3.11) extend to all non-negative integers m, n.

For $T = \text{Min}$ and $n \geq 1$, $x^n = x$; for $T = P$, x^n is the ordinary n-th power of x; and for $T = W$, $x^n = \max(nx - n + 1, 0)$. (When there is a possibility of confusion, we write x_T^n rather than merely x^n.)

For any binary operation B on I, let $B^* : I^2 \to I$ be the function defined by

$$B^*(x, y) = 1 - B(1 - x, 1 - y). \tag{1.3.12}$$

It is evident that B^* is non-decreasing in each place, commutative, or associative if and only if B has the corresponding property, and that $B^{**} = B$. Moreover the graphs of B and B^* are reflections of each other in the point $(\frac{1}{2}, \frac{1}{2}, \frac{1}{2})$.

For any t-norm T we have that $T^*(x, 0) = x$ and $T^*(x, 1) = 1$. Thus the association (1.3.12) yields a dual class of semigroup operations on I, having identity 0 and null element 1. This motivates

Definition 1.3.4 An **s-norm** is a two-place function $S : I^2 \to I$ which satisfies the monotonicity, commutativity, and associativity conditions (1.3.2), (1.3.3), (1.3.4) and the boundary conditions

$$S(x, 0) = S(0, x) = x, \quad S(x, 1) = S(1, x) = 1. \tag{1.3.13}$$

An s-norm is **strict** if it is continuous on I^2 and strictly increasing in each place on $[0, 1)^2$.

We let $\mathcal{S}, \mathcal{S}_{Co}$, and $\mathcal{S}_{St}$ denote the set of s-norms, continuous s-norms and strict s-norms, respectively.

The s-norms associated with the t-norms listed in (1.3.6) are

$$\begin{aligned} \mathrm{Min}^*(x,y) &= \mathrm{Max}(x,y), \\ P^*(x,y) &= x+y-xy, \\ W^*(x,y) &= \mathrm{Min}(x+y,1), \\ Z^*(x,y) &= \begin{cases} x, \ y=0, \\ y, \ x=0, \\ 1, \text{ otherwise.} \end{cases} \end{aligned} \tag{1.3.14}$$

The graphs of these s-norms are depicted in Figure 1.3.3.

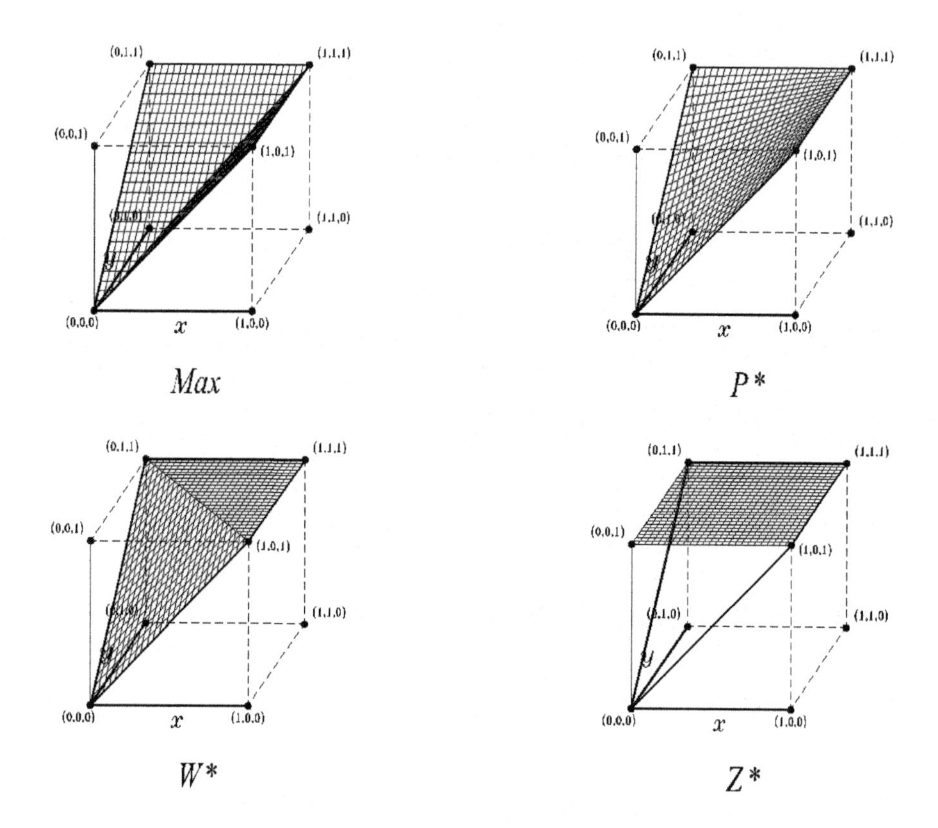

Figure 1.3.3. Graphs of the s-norms Max, P^*, W^*, Z^*.

From the previous discussion it is immediate that T is a t-norm if and only if T^* is an s-norm. It therefore follows that for any statement concerning t-norms there is a corresponding dual statement for s-norms. For instance, T is in $\mathcal{T}_{Co}$ (resp., in $\mathcal{T}_{St}$) if and only if T^* is in $\mathcal{S}_{Co}$ (resp., $\mathcal{S}_{St}$); $T_1 < T_2$ iff $T_1^* > T_2^*$; the s-norms displayed in (1.3.14) satisfy

$$\text{Max} < P^* < W^* < Z^*; \tag{1.3.15}$$

and, for any s-norm S,

$$\text{Max} \le S \le Z^*. \tag{1.3.16}$$

Note also that, since Min < Max,

$$T < S, \tag{1.3.17}$$

for any t-norm T and any s-norm S.

The restrictions of associative functions to the line $y = x$ play a fundamental role in the sequel. Accordingly, the **diagonal** of a t-norm T is the function $\delta_T : I \to I$ defined by

$$\delta_T(x) = T(x, x). \tag{1.3.18}$$

The iterates of δ_T are the functions δ_T^n defined recursively by

$$\delta_T^0 = j_I \quad \text{and} \quad \delta_T^{n+1} = \delta_T \circ \delta_T^n.$$

Lemma 1.3.5 *The diagonal δ_T of the t-norm T has the following properties:*

(a) δ_T is non-decreasing, with $\delta_T(0) = 0$ and $\delta_T(1) = 1$;
(b) $\delta_T(x) \le x$, for all x in I; (1.3.19)
(c) for any x in (0,1), the sequence $\{\delta_T^n(x)\}$ is non-increasing and $0 \le \lim_{n\to\infty} \delta_T^n(x) < 1$;
(d) in terms of the T-powers defined in (1.3.9),

$$\delta_T^n(x) = x^{2^n}. \tag{1.3.20}$$

Proof. Properties (a) and (b) are obvious; (c) now follows from the inequalities $\delta_T^{n+1}(x) = \delta_T(\delta_T^n(x)) \le \delta_T^n(x)$ and $0 \le \delta_T^n(x) \le x$; and a simple induction yields (d). □

An **idempotent** of a t-norm T is an element x of I for which $T(x, x) = x$, i.e., $\delta_T(x) = x$. The elements 0 and 1 are idempotents of every t-norm.

Note that if x is an idempotent of T, then for every $y \geq x$, $x = T(x, x) \leq T(x, y) \leq T(x, 1) = x$, i.e., $T(x, y) = T(y, x) = x$, which immediately yields

Lemma 1.3.6 *If $\delta_T(x) = x$ for all x in I, then $T = Min$.*

At the other extreme, the t-norms for which 0 and 1 are the only idempotents are of special importance. Clearly, for any such T,

$$\delta_T(x) < x, \qquad \text{for } 0 < x < 1. \tag{1.3.21}$$

In particular, every T in $\mathcal{T}_{St}$ satisfies (1.3.21) since, in this case, $\delta_T(x) = T(x, x) < T(x, 1) = x$.

Definition 1.3.7 A t-norm T is **Archimedean** if, for any x, y in (0,1), there exists a positive integer m such that $x^m < y$.

Combining Lemma 1.3.5 with the fact that the sequence $\{x^m\}$ is non-increasing immediately yields

Lemma 1.3.8 *A t-norm T is Archimedean if and only if*

$$\lim_{n\to\infty} \delta_T^n(x) = 0, \qquad \textit{for } 0 < x < 1. \tag{1.3.22}$$

We let $\mathcal{T}_{Ar}$ denote the set of Archimedean elements of $\mathcal{T}_{Co}$.

Theorem 1.3.9 *A continuous t-norm is Archimedean if and only if it does not have any interior idempotents, i.e., if and only if it satisfies (1.3.21). In particular, every strict t-norm is Archimedean, i.e., $\mathcal{T}_{St} \subset \mathcal{T}_{Ar}$.*

Proof. It is sufficient to show that (1.3.21) and (1.3.22) are equivalent when the diagonal δ is continuous. First, if $\delta(x) = x$ for some x in (0,1), then $\delta^n(x) = x$ for all n, contradicting (1.3.22). To prove the converse, observe that $\delta(\lim_{n\to\infty} \delta^n(x)) = \lim_{n\to\infty} \delta(\delta^n(x)) = \lim_{n\to\infty} \delta^n(x)$ for all x, whence (1.3.21) and Lemma 1.3.5(c) together yield (1.3.22). □

For discontinuous t-norms, the Archimedean property is stronger than (1.3.21); moreover, even strict monotonicity of a t-norm does not imply the Archimedean property. A full discussion of these facts, including the extension of Theorem 1.3.9 to all t-norms, is presented in Section 2.5. Note that Z and W are Archimedean but not strictly increasing.

The preceding statements concerning diagonals of t-norms and the Archimedean property have their counterparts for s-norms. These all follow readily from the fact that, for any s-norm S,

$$\delta_S(x) = 1 - \delta_{S^*}(1 - x). \tag{1.3.23}$$

In particular, for s-norms the inequalities (1.3.19) and (1.3.21) are reversed, the sequence δ_S^n is non-decreasing, the Archimedean inequality is $x^m > y$ (where the x^m are S-powers), and the limit in (1.3.22) is 1. Theorem 1.3.9 is valid verbatim for s-norms, upon replacing the inequality in (1.3.21) by $\delta_S(x) > x$.

The **sections** of a t-norm T are the functions obtained by fixing T in one place. Thus, for each y in I, the y-**section** σ_y is the function defined on I by

$$\sigma_y(x) = T(x, y). \tag{1.3.24}$$

In view of the commutativity, the x-sections of T are the same family of functions. Each y-section is non-decreasing, with $\sigma_y(0) = 0$ and $\sigma_y(1) = y$.

1.4 Copulas

A class of binary operations, which are related to t-norms and which are of great importance in probability and statistics, are the copulas introduced by A. Sklar [1959].

Definition 1.4.1 A (two-dimensional) **copula** is a two-place function $C : I^2 \to I$ which satisfies the boundary conditions (1.3.1) and the monotonicity condition

$$C(x_1, y_1) - C(x_2, y_1) - C(x_1, y_2) + C(x_2, y_2) \geq 0, \tag{1.4.1}$$

whenever $x_1 \leq x_2$, $y_1 \leq y_2$.

It is easy to check that W, P and Min satisfy (1.4.1) and are thus copulas. Several basic properties of copulas are collected in

Lemma 1.4.2 *If C is a copula, then*

$$0 \leq C(x_2, y_2) - C(x_1, y_1) \leq x_2 - x_1 + y_2 - y_1, \tag{1.4.2}$$

whenever $x_1 \leq x_2$, $y_1 \leq y_2$. *Thus every copula is non-decreasing in each place and Lipschitz continuous. Moreover, for any copula* C,

$$W \leq C \leq Min. \tag{1.4.3}$$

Proof. To obtain the first inequality in (1.4.2), let $x_1 = 0$ in (1.4.1); let $y_1 = 0$ in (1.4.1) and then relabel y_2 as y_1; add the resulting inequalities. Similarly, to obtain the second, let $x_2 = 1$ in (1.4.1), let $y_2 = 1$ in (1.4.1), then relabel y_1 as y_2 and add these inequalities. Finally, letting $x_2 = y_2 = 1$ in (1.4.1), and combining the result with Lemma 1.3.3 yields (1.4.3). □

In the statistical literature, the largest and smallest copulas, Min and W, in (1.4.3) are generally referred to as the **Fréchet-Hoeffding bounds** [Fréchet (1935), (1951); Hoeffding (1940)].

Copulas were so-named because they link bivariate probability distributions to their margins. The exact connection is given by the following basic result, first announced in [Sklar (1959)]; a proof is given in [Schweizer and Sklar (1974)] (see also [Sklar (1996a); Nelsen (1999)]).

Theorem 1.4.3 *Let* H *be a bivariate probability distribution with margins* F *and* G. *Then there is a copula* C *such that*

$$H(u, v) = C(F(u), G(v)), \tag{1.4.4}$$

for all u, v *in* $\mathbb{R}$. *If* F *and* G *are continuous,* C *is unique; otherwise,* C *is uniquely determined on* $(Ran\ F) \times (Ran\ G)$. *In the other direction, for any univariate distributions* F, G *and any copula* C, *the function* H *defined by* (1.4.4) *is a bivariate distribution with margins* F *and* G.

Copulas therefore provide a natural setting for the study of properties of distributions with fixed margins. Note, in particular, that a copula is itself a continuous bivariate distribution on I^2 with uniform margins. The extension of the copula notion and Theorem 1.4.3 to higher dimensions, also due to A. Sklar [1959], will be presented in Section 4.4.

In the next theorem, we collect the key results which establish the role of copulas in the theory of probability and in mathematical statistics. For the sake of brevity, we restrict our attention to continuous distributions.

Let X and Y be random variables, defined on a common probability space and taking values in $\mathbb{R}$, with continuous distributions F_X, F_Y and joint distribution H_{XY}, i.e.,

$$F_X(u) = Pr(X \leq u), \qquad F_Y(v) = Pr(Y \leq v),$$

$$H_{XY}(u,v) = Pr(X \le u, Y \le v);$$

and denote by C_{XY} the (unique) copula guaranteed by Theorem 1.4.3, so that

$$H_{XY}(u,v) = C_{XY}(F_X(u), F_Y(v)). \tag{1.4.5}$$

Theorem 1.4.4 *Let X, Y, F_X, F_Y, H_{XY}, and C_{XY} be as in the preceding paragraph. Then:*

(a) X and Y are independent if and only if $C_{XY} = P$.
(b) $Y = f(X)$ a.s., where f is strictly increasing a.s. on Ran X, if and only if $C_{XY} = Min$.
(c) $Y = f(X)$ a.s., where f is strictly decreasing a.s. on Ran X, if and only if $C_{XY} = W$.
(d) If f and g are strictly increasing a.s. on Ran X and Ran Y, respectively, then

$$C_{f(X),g(Y)} = C_{XY}.$$

(e) If f and g are strictly decreasing a.s. on Ran X and Ran Y, respectively, then

$$C_{f(X),Y}(x,y) = y - C_{XY}(1-x,y), \tag{1.4.6a}$$
$$C_{X,g(Y)}(x,y) = x - C_{XY}(x,1-y), \tag{1.4.6b}$$
$$C_{f(X),g(Y)}(x,y) = x + y - 1 + C_{XY}(1-x,1-y). \tag{1.4.7}$$

(f) C_{XY} is the restriction to I^2 of the joint distribution of the probability transforms $F_X(X)$ and $F_Y(Y)$.
(g) For continuous distributions F and G, let $X_1 = F^{-1}(F_X(X))$ and $Y_1 = G^{-1}(F_Y(Y))$. Then $F_{X_1} = F$, $F_{Y_1} = G$, and $C_{X_1Y_1} = C_{XY}$.

Parts (a)-(c) are classical results of M. Fréchet [1951, 1957, 1958]. The proofs of the remaining parts are straightforward. These two theorems show that much of the study of joint distributions can be reduced to the study of copulas. In particular, (1.4.5) and part (d) of Theorem 1.4.4 imply that it is precisely the copula that captures those properties of the joint distribution which are invariant under a.s. strictly increasing transformations. Thus, for example, the quantity

$$\sigma(X,Y) = 12 \int_0^1 \int_0^1 |C_{XY}(x,y) - xy| dx dy$$

is a measure of monotone dependence of the random variables X and Y [Schweizer and Wolff (1981)]. Moreover, many well-established stochastic notions are equivalent to simple properties of copulas. For instance, the notion of stochastic dominance translates to pointwise order: If $H_1 = H_{X_1Y_1}$ and $H_2 = H_{X_2Y_2}$ are bivariate distributions with equal margins, i.e., $F_{X_1} = F_{X_2}$ and $F_{Y_1} = F_{Y_2}$, then (X_2, Y_2) is said to dominate (X_1, Y_1) if $H_2 \geq H_1$, which is the case precisely when their copulas satisfy $C_2 \geq C_1$.

In the years between the appearance of the seminal paper by A. Sklar [1959] and the papers by the third author and E.F. Wolff [1976, 1981], most of the results concerning copulas were obtained in connection with problems arising from the study of probabilistic metric spaces. Then, in the decade 1980-1990, research on copulas and their applications grew markedly and culminated in an international conference which was held in Rome in 1990. Since then, four additional international conferences have been held – in Seattle in 1993, in Prague in 1996, in Barcelona in 2000 and in Quebec City in 2004; and since the turn of the century there has been an explosion of copula-centered activity, fueled largely by the significant role that these functions play in the area of finance and risk management. This monograph is not the place to review the extensive and rapidly expanding literature. Instead, we refer the reader to the proceedings of the above-mentioned conferences [Dall'Aglio, Kotz and Salinetti (1991); Rüschendorf, Schweizer and Taylor (1996); Beneš and Štěpán (1997); Cuadras, Fortiana and Rodriguez-Lallena (2002); Genest et al. (2005)] for details; to the paper [Schweizer (1991)] for a survey of developments from 1959 to 1989; to the papers [Dall'Aglio (1991)] and [Sklar (1996a)] for interesting comments on the early history of the subject; to the monograph [Nelsen (1999)] for a comprehensive and eminently readable introduction to the subject, together with an extensive bibliography; and to the books [Hutchinson and Lai (1990); Joe (1997); Drouet Mari and Kotz (2001); Cherubini, Luciano and Vecchiato (2004)].

In this monograph we are primarily concerned with associative copulas. A t-norm that satisfies (1.4.1) is a copula, and in view of Lemma 1.4.2 and the paragraph immediately preceding Definition 1.3.2, an associative copula is a t-norm. Many of the important copulas and families of copulas are associative. However, there are commutative copulas that are not associative, and hence not t-norms; and there are t-norms that satisfy (1.4.3) but not (1.4.1), and hence are not copulas. (See Examples A.2.1 and A.2.2 of Appendix A.)

The following characterization of associative copulas is due to R. Moynihan [1978] (see also [Schweizer and Sklar (1983), (2005), Theorem 6.3.2]).

Theorem 1.4.5 *A t-norm T is a copula if and only if it satisfies the Lipschitz condition*

$$T(x_2, y) - T(x_1, y) \leq x_2 - x_1, \qquad \text{whenever } x_1 \leq x_2. \tag{1.4.8}$$

Proof. It is obvious from (1.4.2) that any copula satisfies (1.4.8). In the other direction, if T satisfies (1.4.8), then T is continuous. Choose $x_1 \leq x_2$ and $y_1 \leq y_2$. Since $T(0, y_2) = 0 \leq y_1 \leq y_2 = T(1, y_2)$, there exists a number z in I such that $T(z, y_2) = y_1$. Hence, by (1.3.3), (1.3.4), and (1.4.8),

$$\begin{aligned} T(x_2, y_1) - T(x_1, y_1) &= T(x_2, T(z, y_2)) - T(x_1, T(z, y_2)) \\ &= T(T(x_2, y_2), z) - T(T(x_1, y_2), z) \\ &\leq T(x_2, y_2) - T(x_1, y_2), \end{aligned}$$

which is (1.4.1). □

The **dual copula** of the copula C is the function $C^\wedge$ defined on I^2 by

$$C^\wedge(x, y) = x + y - C(x, y). \tag{1.4.9}$$

Dual copulas are motivated by a natural probabilistic interpretation: in the notation of (1.4.5),

$$\begin{aligned} Pr(X \leq u \text{ or } Y \leq v) &= F_X(u) + F_Y(v) - C_{XY}(F_X(u), F_Y(v)) \\ &= C^\wedge_{XY}(F_X(u), F_Y(v)). \end{aligned} \tag{1.4.10}$$

The properties of $C^\wedge$ listed below follow readily from Lemma 1.4.2.

Lemma 1.4.6 *If C is a copula and $C^\wedge$ is its dual copula, then:*

(a) Ran $C^\wedge = I$, i.e., $C^\wedge$ is a binary operation on I;
(b) $C^\wedge$ satisfies the boundary conditions (1.3.13);
(c) $C^\wedge$ is non-decreasing in each place;
(d) $C^\wedge$ is commutative if and only if C is commutative.

It is emphatically not true in general that $C^\wedge$ is an s-norm when C is a t-norm. This is so because $C^\wedge$ usually fails to be associative. The canonical copulas W, P, and Min, however, are exceptions: for each of these, $C^{\wedge} = C^*$. The complete answer to the question as to when C and $C^{\wedge}$ are simultaneously associative was given by the second author in [Frank (1979)] where he showed that this is the case if and only if C belongs to the important family of copulas that now bears his name. The details are given in Section 3.1.

In [Alsina, Nelsen and Schweizer (1993)] the notion of a copula was extended to that of a **quasi-copula**. The original definition was rather unwieldy. Then, in [Genest, Quesada-Molina, Rodriguez-Lallena and Sempi (1999)] it was proved that $Q : I^2 \to I$ is a quasi-copula if and only if Q satisfies (1.3.1), (1.3.2) and (1.4.8). hence, in view of Theorem 1.4.5, every associative quasi-copula is a copula.

With the discovery of the new definition, the pace of research on quasi-copulas quickened. A brief review of recent results is given in [Nelsen (2005)]. Here we only note that the set of 2-quasi-copulas is the Dedekind-MacNeille completion of the poset of 2-copulas [Nelsen and Úbeda-Flores, to appear].

Finally, we note that the set of copulas is convex.

Chapter 2

Representation theorems for associative functions

2.1 Continuous, Archimedean t-norms

The representation of continuous Archimedean t-norms by one-place functions and addition was discovered by C.-H. Ling [1965] as a variant of the Abel-Aczél representation of strict t-norms ([Aczél (1949), (1966); Schweizer and Sklar (1963)]). Since then, the representation has been established under successively weaker assumptions. In this section we present an especially simple version whose proof can be readily adapted to a variety of extensions (which will be discussed in Section 2.7).

We begin by assuming only that T is associative, is continuous, and satisfies a relaxation of the t-norm boundary conditions. The first step of the proof is to show that these assumptions imply the remaining boundary conditions and monotonicity (compare [Mostert and Shields (1957); Paalman-de Miranda (1964)]).

Lemma 2.1.1 *Suppose that $T : I^2 \to I$ satisfies the following conditions:*

(i) $T(x, 0) = T(0, x) = 0,$ *for all x in I,*
(ii) $T(1, 1) = 1,$
(iii) T is associative,
(iv) T is jointly continuous.

Then

(a) $T(x, 1) = T(1, x) = x,$ *for all x in I,*

and

(b) T is non-decreasing in each place.

Proof. Choose x in I. Since $T(0,1) = 0$, $T(1,1) = 1$, and T is continuous, there exists a z such that $T(z,1) = x$. Thus

$$T(x,1) = T(T(z,1),1) = T(z,T(1,1)) = T(z,1) = x.$$

Similarly, $T(1,x) = x$. This establishes (a).

To prove (b), fix x in I and let

$$x_0 = \sup\{t \text{ in } I : T(u,v) \leq x \text{ for all } (u,v) \text{ in } [0,t] \times I\}.$$

Since $T(0,v) = 0$ for all v, the above set is not empty, so that x_0 is well-defined.

Note first that, in view of (a), $x_0 \leq x$; and note further that, by continuity, $T(x_0, v) \leq x$ for all v in I.

Next, there exists a z in I for which $T(x_0, z) = x$. If $x = 0$, then $x_0 = 0$ and any z in I will do. So, assume that $x > 0$ and suppose, to the contrary, that $T(x_0, v) < x$ for all v in I. Then since T is continuous, and $T(x_0, 0) = 0$, there is an $\epsilon > 0$ such that $T(x_0, v) \leq x - \epsilon$ for all v; and since T is uniformly continuous on I^2, there is an $x' > x_0$ such that $T(x', v) < x$ for all v, contradicting the maximality of x_0. Hence for each y in I, we have

$$T(x,y) = T(T(x_0,z),y) = T(x_0, T(z,y)) \leq x.$$

A similar argument yields $T(x,y) \leq y$, whence it follows that $T(x,y) \leq \text{Min}(x,y)$. Finally, suppose that $0 \leq x < y \leq 1$. Since

$$T(0,y) = 0 \leq x < y = T(1,y),$$

there exists a z such that $T(z,y) = x$. Consequently, since $T \leq \text{Min}$, for each w in I,

$$T(x,w) = T(T(z,y),w) = T(z,T(y,w)) \leq T(y,w).$$

Thus T is non-decreasing in its first place. A similar argument shows that T is non-decreasing in its second place, and the proof is complete. □

An interesting open question is whether part (b) of Lemma 2.1.1 is true when assumption (iv) is weakened to continuity in each place. If the answer is affirmative, then joint continuity of T is a consequence of the following easily established fact [Kruse and Deely (1969)]:

Lemma 2.1.2 *Any function which is continuous in each place and non-decreasing in each place is jointly continuous.*

The following examples show that, without both of the boundary conditions (i) and (ii), part (b) of Lemma 2.1.1 is false.

Example 2.1.3 For (x, y) in I^2, let

$$T_1(x, y) = xy + (1 - x)(1 - y),$$

and

$$T_2(x, y) = 2 \text{ Min}(x, 1 - x) \text{ Min}(y, 1 - y).$$

Both T_1 and T_2 are continuous functions from I^2 to I and, as is readily checked, are associative. Now $T_1(x, 1) = T_1(1, x) = x$, but $T_1(x, 0) = T_1(0, x) = 1 - x$ which shows both that the boundary condition (i) fails and that T_1 fails to be non-decreasing. Next, $T_2(x, 0) = T_2(0, x) = 0$, but $T_2(1, 1) = 0$ and $T_2(x, y) < T_2\left(\frac{1}{2}, \frac{1}{2}\right) = \frac{1}{2}$ for all $(x, y) \neq \left(\frac{1}{2}, \frac{1}{2}\right)$.

We now turn our attention to the basic representation theorem for t-norms which is the focal point of this section and, indeed, the cornerstone of this monograph. We begin with some preliminaries.

Definition 2.1.4 A **right-inverse** of a function g is a function f such that

$$\text{Dom } f = \text{Ran } g, \qquad \text{Ran } f \subseteq \text{Dom } g,$$

and

$$g \circ f = j_{\text{Ran } g}, \text{ i.e., } g(f(x)) = x \text{ for every } x \text{ in Ran } g.$$

It easily follows from the Axiom of Choice that any function whatever has a right-inverse; indeed the converse is also true. An elementary argument shows that if f is a right-inverse of g, then f is invertible and f^{-1} is the restriction of g to Ran f; moreover, f is uniquely determined if and only if g is one-to-one.

Definition 2.1.5 Let $t : I \to \mathbb{R}^+$ be continuous and strictly decreasing, with $t(1) = 0$. The **pseudo-inverse** of t is the function $t^{(-1)}$, for which Dom $t^{(-1)} = \mathbb{R}^+$ and Ran $t^{(-1)} = I$, given by

$$t^{(-1)}(u) = \begin{cases} t^{-1}(u), & 0 \le u \le t(0), \\ 0, & t(0) \le u \le \infty. \end{cases} \tag{2.1.1}$$

Clearly, $t^{(-1)}$ is continuous, non-increasing, and strictly decreasing on $[0, t(0)]$. Since

$$t^{(-1)} \circ t = j_I, \tag{2.1.2}$$

t is a right-inverse of $t^{(-1)}$. Note also that

$$t \circ t^{(-1)}(u) = \begin{cases} u, & 0 \le u \le t(0), \\ t(0), & t(0) \le u \le \infty, \end{cases}$$

$$= \min(u, t(0)). \tag{2.1.3}$$

If t is onto $\mathbb{R}^+$, i.e., if $t(0) = \infty$, then $t^{(-1)} = t^{-1}$, and conversely.

Theorem 2.1.6 *(Representation Theorem) Suppose that $T : I^2 \to I$ satisfies the following conditions:*

(i) $T(x, 0) = T(0, x) = 0$, *for all x in I,*
(ii) $T(1, 1) = 1$,
(iii) T is associative,
(iv) T is jointly continuous,
(v) T is Archimedean, i.e., for all x, y in $(0, 1)$, there exists a positive integer n such that $x^n < y$.

Then T admits the representation

$$T(x, y) = t^{(-1)}(t(x) + t(y)), \tag{2.1.4}$$

where t is a continuous and strictly decreasing function from I to $\mathbb{R}^+$, with $t(1) = 0$, and $t^{(-1)}$ is the pseudo-inverse of t.

Proof. It has been established that T also fulfills the following conditions:

(vi) $T(x, 1) = T(1, x) = x$, for all x in I,
(vii) T is non-decreasing in each place,
(viii) $T(x, y) \le \text{Min}(x, y)$,
(ix) $T(x, x) < x$, for $0 < x < 1$.

(See Lemma 2.1.1, Lemma 1.3.3, and Theorem 1.3.9.)

Whenever convenient, we write xy for $T(x, y)$; we define the sequence of functions $\{f_n\}$, $n \ge 1$, by

$$f_n(x) = x^n,$$

where the x^n are the T-powers defined in (1.3.9); and, omitting no details, we break the long argument into a sequence of shorter steps.

(1) For all $n \geq 1$, f_n is continuous and non-decreasing on I, with $f_n(0) = 0$ and $f_n(1) = 1$, whence Ran $f_n = I$.

Proof. These properties are either immediate or easy inductions.

(2) For each x in I, the sequence $\{x^n\}$ is non-increasing, i.e., $f_{n+1}(x) \leq f_n(x)$, $n = 1, 2, \ldots$.

Proof. $x^{n+1} = xx^n \leq 1x^n = x^n$.

(3) For each x in $[0, 1)$, $\lim_{n\to\infty} x^n = 0$, i.e., $\lim_{n\to\infty} f_n(x) = 0$.

Proof. If $\lim_{n\to\infty} x^n = y > 0$, then $x^n \geq y$ for all n, which contradicts (v).

(4) For each (x, y) in $(0, 1)^2$, $xy < y$ and $xy < x$, i.e.,

$$T(x, y) < \text{Min}(x, y).$$

Proof. By (viii), $xy \leq y$. Suppose $xy = y$. Then by induction, $x^n y = y$ for all n. Consequently, from (iv) and (3),

$$y = \lim_{n\to\infty}(x^n y) = (\lim_{n\to\infty} x^n)y = 0y = 0,$$

which is a contradiction. Thus $xy < y$; and similarly $xy < x$.

(5) For each x in (0,1), if $x^n > 0$, then $x^{n+1} < x^n$, i.e., if $0 < f_n(x) < 1$, then $f_{n+1}(x) < f_n(x)$.

Proof. Let $y = x^n$ in (4).

(6) If $0 \leq y < x \leq 1$, then there exist z, z' such that $y = xz = z'x$.

Proof. This follows immediately from continuity and the inequalities $x0 = 0 \leq y < x = x1$, and $0x \leq y < 1x$.

(7) If $x^n > 0$, then $y^n < x^n$ for all $y < x$, i.e., if $f_n(x) > 0$, then $f_n(y) < f_n(x)$ for all $y < x$.

Proof. The statement is true if $y = 0$ trivially, or if $x = 1$ by (2). Suppose now that (7) is false. Then there exist x, y such that $0 < y < x < 1$, $x^n > 0$, and $y^n \geq x^n$. By (6) there is a $z < 1$ for which $y = xz$. Consequently, using (1), (vii) and (4), we obtain

$$0 < x^n \leq y^n = y^{n-1}y = y^{n-1}xz \leq x^{n-1}xz = x^n z < x^n,$$

which is a contradiction.

To this point we have shown that the graphs of the functions f_n have no "interior flat pieces", i.e., that for each n, as x decreases from 1 to 0, $f_n(x)$ decreases steadily from 1 until it attains the value 0, from which point on it remains constant. Moreover, these graphs are strictly ordered in the interior of the unit square.

For example, if $T = P$, then $f_n = j^n$, the n-th power function; and if $T = W$, then $f_n(x) = \max(nx - n + 1, 0)$. These sequences are depicted in Figure 2.1.1.

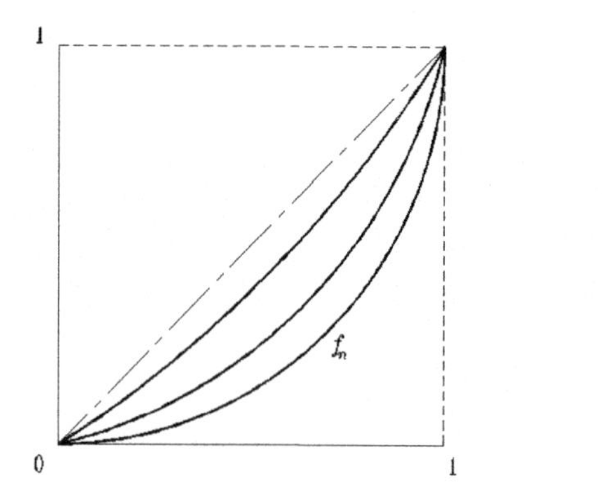

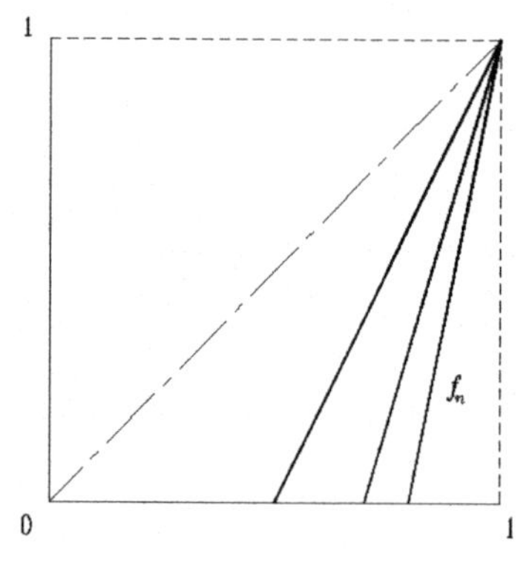

Figure 2.1.1. The functions f_n for $T = P$ and $T = W$.

(8) Suppose there is an x_0 in (0,1) that is **nilpotent**, i.e., such that $x_0^N = 0$ for some $N > 1$. Then every element of (0,1) is nilpotent.

Proof. If $y < x_0$, then $0 \leq y^N \leq x_0^N = 0$, whence $y^N = 0$. If $y > x_0$, then $y^m < x_0$ for some m by (v), and so $0 \leq y^{mN} \leq x_0^N = 0$, whence $y^{mN} = 0$.

(9) Let $\alpha_n = \sup\{x : f_n(x) = 0\}$. Then for all $n \geq 1$, $0 \leq \alpha_n < 1$, $\alpha_n \leq \alpha_{n+1}$, $f_n(\alpha_n) = 0$, and f_n is strictly increasing on $[\alpha_n, 1]$. Moreover, if $\alpha_2 > 0$, then $\lim_{n\to\infty} \alpha_n = 1$; and if $\alpha_2 = 0$, then $\alpha_n = 0$ for all n.

Proof. All of the properties listed in the second sentence follow immediately from previous statements. Suppose next that $\alpha_2 > 0$ and choose any $x < 1$. Since $\alpha_2^2 = f_2(\alpha_2) = 0$, α_2 is nilpotent, whence by (8) $x^N = 0$ for some N. But then $\alpha_N \geq x$, and so $\alpha_n \geq x$ for all $n \geq N$. Hence $\lim_{n\to\infty} \alpha_n = 1$. Lastly suppose that $\alpha_2 = 0$ and that there is an integer $n > 2$ such that $\alpha_n > 0$. Let N be the smallest such integer, and let $\alpha_N = x > 0$. If $N = 2k$, then since $0 = x^N = (x^k)^2$ and $\alpha_2 = 0$, we must have $x^k = 0$; but then $\alpha_k \geq x > 0$ and $k < N$, which is a contradiction. If $N = 2k + 1$, we find that $\alpha_{k+1} \geq x > 0$, which is again a contradiction.

According to (9), if the diagonal $\delta = f_2$ is strictly increasing on I, then so are all the f_n; if not, then the lengths of the intervals on which f_n is strictly increasing shrink to 0 as n increases. Thus there are two possible modes of behavior of the sequence $\{f_n\}$, illustrated by Figure 2.1.1.

(10) Let f_n^* be the unique right-inverse of f_n for which $f_n^*(0) = \alpha_n$. Then:
(a) Dom $f_n^* = I$, Ran $f_n^* = [\alpha_n, 1]$.
(b) f_n^* is continuous and strictly increasing, and $f_n \circ f_n^* = j_I$.
(c) $f_n^* \circ f_n(x) = \begin{cases} \alpha_n, & 0 \leq x \leq \alpha_n, \\ x, & \alpha_n \leq x \leq 1. \end{cases}$
(d) f_n^* is the inverse of the restriction of f_n to $[\alpha_n, 1]$.

(11) For each x in (0,1) and all n, $f_{n+1}^*(x) > f_n^*(x)$.

Proof. If not, then there is an x in (0,1) and an integer n such that

$$0 \leq \alpha_n \leq \alpha_{n+1} < f_{n+1}^*(x) \leq f_n^*(x) < 1.$$

Consequently,

$$0 < f_n \circ f_{n+1}^*(x) \leq f_n \circ f_n^*(x) = x,$$

whence, by (5),

$$x = f_{n+1} \circ f_{n+1}^*(x) < f_n \circ f_{n+1}^*(x) \leq x,$$

which is a contradiction.

(12) For each x in $(0,1)$, $\lim_{n\to\infty} f_n^*(x) = 1$.

Proof. If $\lim_{n\to\infty} f_n^*(x) = y < 1$, then by (11), $f_n^*(x) < y$ for all n, whence $x = f_n \circ f_n^*(x) \leq f_n(y) = y^n$ for all n, which contradicts (v).

The functions f_n^* are, of course, the mirror images in the diagonal of the unit square of the increasing portions of the functions f_n. For example, adding the graphs of the appropriate inverses to Figure 2.1.1 yields Figure 2.1.2.

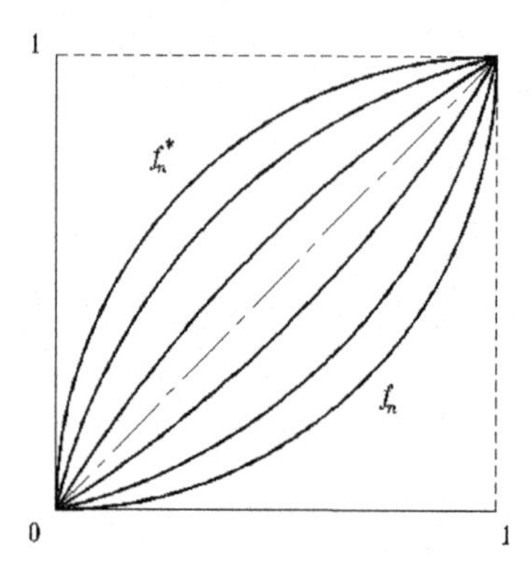

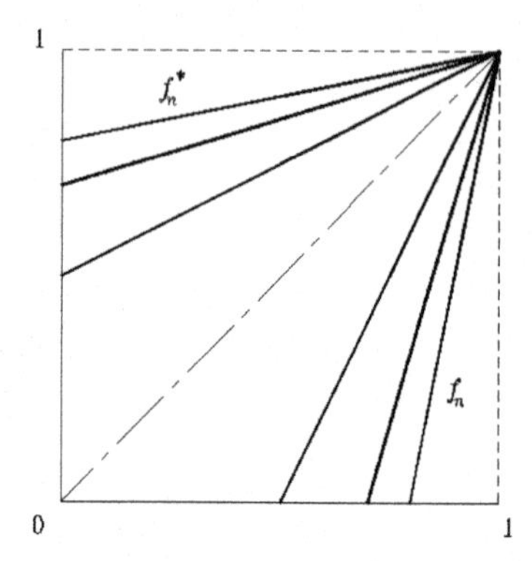

Figure 2.1.2. The functions f_n and f_n^* for $T = P$ and $T = W$.

Note that the laws of exponents for T-powers, namely (1.3.10) and (1.3.11), can be written as follows: for all $m, n \geq 1$,

$$f_{m+n} = T(f_m, f_n) = T(f_n, f_m),$$

and

$$f_{mn} = f_n \circ f_m = f_m \circ f_n.$$

(13) For all $m, n \geq 1$, $f_{mn}^* = f_m^* \circ f_n^* = f_n^* \circ f_m^*$.

Proof. By the preceding display and (10b),

$$f_{mn} \circ f_m^* \circ f_n^* = f_n \circ f_m \circ f_m^* \circ f_n^* = j_I.$$

Thus both f_{mn}^* and $f_m^* \circ f_n^*$ are right-inverses of f_{mn} and, in view of (10d), inverses of the restriction of f_{mn} to $[\alpha_{mn}, 1]$. Hence they are equal, and, by a similar argument, equal to $f_n^* \circ f_m^*$ as well.

(14) For all $k, m, n \geq 1$, $f_m \circ f_n^* = f_{km} \circ f_{kn}^*$.

Proof. By (10b) and (13),

$$f_m \circ f_n^* = f_m \circ f_k \circ f_k^* \circ f_n^* = f_{km} \circ f_{kn}^*.$$

Let Q^+ be the set of positive rationals. For $r = m/n$ in Q^+, define the function $f_r : I \to I$ by

$$f_r = f_m \circ f_n^*.$$

In view of (14) and the fact that $f_1^* = j_I$, the functions f_r are well-defined for all r in Q^+. Clearly each f_r is continuous and non-decreasing.

(15) For each x in I and r, s in Q^+, if $r < s$ and $f_r(x) > 0$, then $f_r(x) > f_s(x)$.

Proof. Let $r = m/d$ and $s = n/d$, so that $m < n$. Then from (5) and (2),

$$f_r(x) = f_m \circ f_d^*(x) > f_{m+1} \circ f_d^*(x) \geq f_n \circ f_d^*(x) = f_s(x).$$

Now fix c in (0,1) and define the function $g : Q^+ \to I$ by

$$g(r) = f_r(c).$$

The preceding discussion then yields:

(16) The function g is non-increasing and is strictly decreasing whenever it is positive.

The graphs of the functions f_r "fill in" the regions between the graphs depicted in Figure 2.1.2 in an "orderly" fashion. The function g may therefore be viewed as the vertical "cross-section" obtained by evaluating the functions f_r at c. (See Figure 2.1.3.)

(17) For all r, s in Q^+, $g(r+s) = T(g(r), g(s))$.

Proof. Let $r = m/d$ and $s = n/d$. Then

$$g((m+n)/d) = f_{m+n} \circ f_d^*(c) = T(f_m \circ f_d^*(c), f_n \circ f_d^*(c)) = T(g(r), g(s)).$$

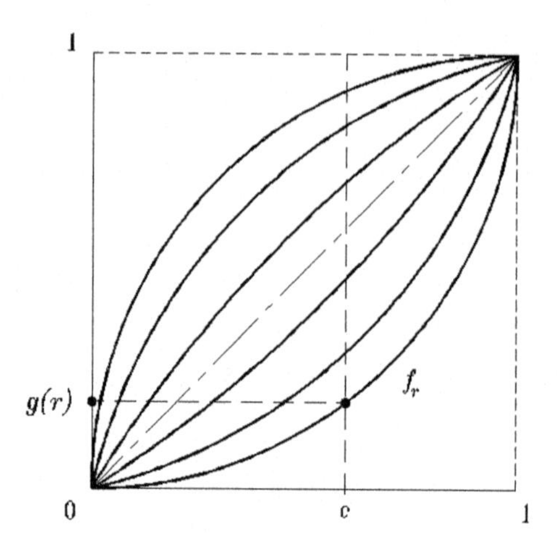

Figure 2.1.3. The functions f_r and g for $T = P$.

(18) $\lim_{n\to\infty} g\left(\frac{1}{n}\right) = 1$.

Proof. This is a restatement of (12) since $g\left(\frac{1}{n}\right) = f_n^*(c)$.

Accordingly, we define $g(0) = 1$ and note that (16) and (17) remain valid on $Q^+ \cup \{0\}$.

(19) The function g is uniformly continuous on $Q^+ \cup \{0\}$.

Proof. Let $\epsilon > 0$ be given. Since T is uniformly continuous on I^2, there is a $\delta > 0$ such that $|T(x, y) - T(u, v)| < \epsilon$ whenever $|x - u| < \delta$ and $|y - v| < \delta$. Choose h in Q^+ so that $0 < 1 - g(h) < \delta$. Then for any r in $Q^+ \cup \{0\}$, (16) and (17) yield

$$0 \leq g(r) - g(r + h) = T(g(r), 1) - T(g(r), g(h)) < \epsilon.$$

Next, if $r \geq h$, then

$$\begin{aligned} 0 \leq g(r - h) - g(r) &= g(r - h) - g(r - h + h) \\ &= T(g(r - h), 1) - T(g(r - h), g(h)) < \epsilon; \end{aligned}$$

whereas if $r < h$, then $g(r) \geq g(h)$, so that $1 - g(r) < \delta$, whence

$$g(r-h) - g(r) \leq 1 - g(r) = T(1,1) - T(1, g(r)) < \epsilon.$$

Thus, for all r, $g(r-h) - \epsilon \leq g(r) \leq g(r+h) + \epsilon$, whenever h is sufficiently small.

We now complete the proof of the representation theorem by constructing the function t in (2.1.4) from the function g. By (19), g admits an extension $\overline{g}$ to $[0, \infty)$ which is continuous and non-increasing, with $\overline{g}(0) = 1$. Moreover, $\overline{g}$ decreases steadily from 1 unless and until it attains the value 0, after which it is constant. We may therefore, without loss of consistency, define $\overline{g}(\infty) = 0$. Since T is continuous, (17) extends to

$$\overline{g}(u+v) = T(\overline{g}(u), \overline{g}(v)), \tag{2.1.5}$$

for all u, v in $\mathbb{R}^+$.

Let $u_0 = \inf\{u : \overline{g}(u) = 0\}$, and note that $0 < u_0 \leq \infty$. Let t be the unique right-inverse of $\overline{g}$ for which $t(0) = u_0$. Then Dom $t = I$, Ran $t = [0, u_0]$, t is continuous and strictly decreasing, and $t(1) = 0$. Moreover, $\overline{g} = t^{(-1)}$, the pseudo-inverse of t.

Finally, for x, y in I, let $t(x) = u$, $t(y) = v$. Then since we have $\overline{g}(u) = t^{(-1)} \circ t(x) = x$ and $\overline{g}(v) = y$, (2.1.5) transforms into (2.1.4), and the proof is complete. □

Note that the representation (2.1.4) can be written in the equivalent form

$$t(T(x,y)) = \min[t(x) + t(y), t(0)]. \tag{2.1.6}$$

The representation and the commutativity of addition instantly yield

Corollary 2.1.7 *If T satisfies the hypotheses of Theorem 2.1.6, then T is commutative and thus T is a t-norm; specifically, T is an element of $\mathcal{T}_{Ar}$.*

The converse of Theorem 2.1.6 is readily established.

Theorem 2.1.8 *Let $t : I \to \mathbb{R}^+$ be continuous and strictly decreasing, with $t(1) = 0$, and let $t^{(-1)}$ be its pseudo-inverse. Then the function T defined on I^2 by (2.1.4) is a continuous Archimedean t-norm, i.e., T is in $\mathcal{T}_{Ar}$.*

Proof. Since Ran $t^{(-1)} = I$, $T(x,y)$ is in I for every (x,y) in I^2. The boundary conditions (1.3.1) follow readily from (2.1.1). Since t and $t^{(-1)}$

are non-increasing, T is non-decreasing in each place; and since t, $t^{(-1)}$, and addition are continuous, so is T. Next, T is Archimedean because, for any x in (0,1), we have

$$\begin{aligned} T(x,x) &= t^{(-1)}(2t(x)) = t^{(-1)}[\min(2t(x), t(0))] \\ &< t^{(-1)}(t(x)) = x. \end{aligned}$$

It remains to verify associativity. In view of (2.1.6), we have

$$\begin{aligned} t(T(T(x,y),z)) &= \min[t(T(x,y)) + t(z), t(0)] \\ &= \min[t(x) + t(y) + t(z), t(0)] \\ &= t(T(x, T(y,z))). \end{aligned}$$

Lastly, an application of $t^{(-1)}$ completes the proof. □

Motivated by the representation theorem and its converse, we say that the Archimedean t-norm T is **additively generated** by the functions t and $t^{(-1)}$ given in (2.1.4); we call t an **inner additive generator** of T and $t^{(-1)}$ an **outer additive generator** of T.

It follows from Theorem 2.1.8 that Archimedean t-norms are as easy to construct as additive generators. Section 2.6 will be devoted to a large number of examples. Here we point out only that additive generators of W and P are given, respectively, by

$$t_W(x) = 1 - x, \qquad t_W^{(-1)}(u) = \max(1-u, 0) \tag{2.1.7}$$

and

$$t_P(x) = -\log x, \qquad t_P^{-1}(u) = e^{-u}. \tag{2.1.8}$$

For T in $\mathcal{T}_{Ar}$, let $\mathcal{Z}(T)$ denote the **zero set** of T, i.e.,

$$\mathcal{Z}(T) = \{(x,y) \text{ in } I^2 : T(x,y) = 0\}. \tag{2.1.9}$$

Theorem 2.1.9 *Let T be an element of $\mathcal{T}_{Ar}$, and let t be an inner additive generator of T. Then:*

(a) T is strictly increasing in each place on $I^2 \backslash \mathcal{Z}(T)$.
(b) T is strict, i.e., in $\mathcal{T}_{St}$, if and only if $t(0) = \infty$, or equivalently, if and only if $t^{(-1)}$ is strictly decreasing on $\mathbb{R}^+$. In this case

$$T(x,y) = t^{-1}(t(x) + t(y)). \tag{2.1.10}$$

(c) T is strict if and only if there are no interior nilpotent elements.

Proof. From (2.1.1) and (2.1.4) it is immediate that (x, y) is in $\mathcal{Z}(T)$ if and only if $t(x) + t(y) \geq t(0)$. Choose $(x_1, y), (x_2, y)$ in $I^2 \backslash \mathcal{Z}(T)$, with $x_1 < x_2$. Then $t(x_2) + t(y) < t(x_1) + t(y) < t(0)$. Now since $t^{(-1)}$ is strictly decreasing on $[0, t(0)]$, we have

$$T(x_1, y) = t^{(-1)}(t(x_1) + t(y)) < t^{(-1)}(t(x_2) + t(y)) = T(x_2, y),$$

which establishes (a).

If $t(0) = \infty$, so that $t^{(-1)} = t^{-1}$, then (x, y) is in $\mathcal{Z}(T)$ if and only if $x = 0$ or $y = 0$. Thus, by (a), T is strict. Conversely, suppose that $t(0) = c < \infty$, and let $x_0 = t^{(-1)}(c/2)$. Then $0 < x_0 < 1$ and $T(x_0, x_0) = t^{(-1)}(c/2 + c/2) = 0$, whence T is not strict, completing the proof of (b). (See Figure 2.1.4.)

Finally, (c) follows from (b): if T is strict, then $x^n = t^{-1}(nt(x)) > 0$ for any $x > 0$; otherwise, the point x_0 in the preceding paragraph is nilpotent since $x_0^2 = 0$. □

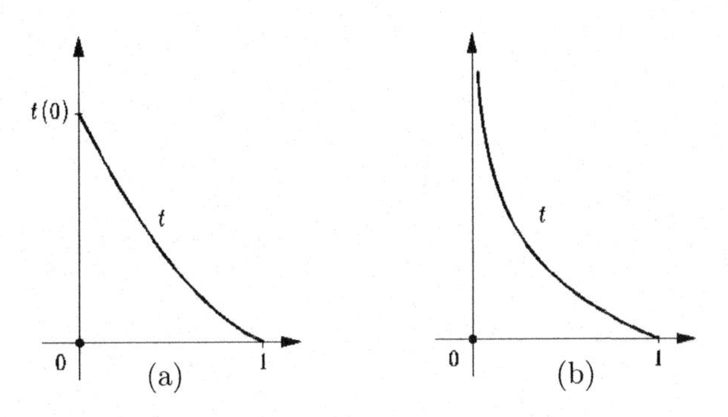

Figure 2.1.4. Inner additive generators of continuous Archimedean t-norms: (a) non-strict, (b) strict.

When T belongs to $\mathcal{T}_{Ar} \backslash \mathcal{T}_{St}$, the boundary of $\mathcal{Z}(T)$ consists of the two line segments $\{0\} \times I$ and $I \times \{0\}$ and the curve determined by the equation

$$t(x) + t(y) = t(0) \tag{2.1.11}$$

or, equivalently,

$$y = \beta_T(x) = t^{(-1)}(t(0) - t(x)). \tag{2.1.12}$$

We shall call the graph of β_T the **boundary curve** of $\mathcal{Z}(T)$. Note that it joins the points (0,1) and (1,0), that it is symmetric with respect to the line $y = x$, and that β_T is strictly decreasing.

Two semigroups are said to be **iseomorphic** if there is a mapping between them which is both an algebraic isomorphism and a topological homeomorphism.

The following characterization of elements of $\mathcal{T}_{Ar}$, first obtained in [Faucett (1955)], is a simple consequence of the preceding results.

Theorem 2.1.10 *Let T be an element of $\mathcal{T}_{Ar}$. If T is strict, then (I,T) is iseomorphic to (I,P); if T is not strict, then (I,T) is iseomorphic to (I,W). More precisely, suppose that t is an inner additive generator of T. Then the order-preserving iseomorphism ϕ is given by*

$$\phi(x) = e^{-t(x)}, \qquad \text{for } T \text{ in } \mathcal{T}_{St},$$

or

$$\phi(x) = 1 - t(x)/t(0), \qquad \text{for } T \text{ in } \mathcal{T}_{Ar}\backslash\mathcal{T}_{St}.$$

Proof. It is clear that, in each case, ϕ is continuous and strictly increasing from I onto I. In the strict case, $t(T(x,y)) = t(x) + t(y)$ by (2.1.10), and so

$$\phi(T(x,y)) = e^{-[t(x)+t(y)]} = \phi(x)\phi(y) = P(\phi(x),\phi(y)).$$

In the non-strict case, $t(T(x,y))/t(0) = \min[(t(x)+t(y))/t(0), 1]$ by (2.1.6), and so

$$\phi(T(x,y)) = \max[\phi(x)+\phi(y)-1, 0] = W(\phi(x),\phi(y)). \qquad \square$$

Lemma 2.1.11 *Let T_1 be a continuous Archimedean t-norm with inner additive generator t_1, let ϵ in $(0,1)$ be given, and let T_2 be a continuous Archimedean t-norm having an inner additive generator t_2 that agrees with t_1 on the interval $[\epsilon,1]$. Then, for all x,y in I,*

$$|T_1(x,y) - T_2(x,y)| \le \epsilon.$$

Proof. Clearly, $t_1(\epsilon) \ge t_1(x) + t_1(y)$ if and only if $t_2(\epsilon) \ge t_2(x) + t_2(y)$, from which it follows that T_1 and T_2 agree whenever they assume a value greater than or equal to ϵ. $\square$

Choosing T_2 to be a strict t-norm yields:

Theorem 2.1.12 *Every continuous Archimedean t-norm is the uniform limit of strict t-norms.*

The representation theorem for t-norms and its corollaries can be translated to analogous results for s-norms by way of the duality expressed in (1.3.12) and the subsequent discussion. Recall that $S : I^2 \to I$ is an s-norm (which is in $\mathcal{S}_{Co}$, $\mathcal{S}_{Ar}$, or $\mathcal{S}_{St}$) if and only if there is a t-norm T (which is in $\mathcal{T}_{Co}, \mathcal{T}_{Ar}$, or $\mathcal{T}_{St}$) such that

$$S(x,y) = T^*(x,y) = 1 - T(1-x, 1-y). \tag{2.1.13}$$

Notice first that Lemma 2.1.1 immediately yields

Lemma 2.1.13 *Suppose that $S : I^2 \to I$ is associative, is jointly continuous, and satisfies the boundary conditions*

$$S(0,0) = 0, \quad S(x,1) = S(1,x) = 1.$$

Then S is also non-decreasing in each place and satisfies all of the s-norm boundary conditions (1.3.13).

To obtain the representation, suppose that S in $\mathcal{S}_{Ar}$ is given by (2.1.13) and that t generates T. Clearly, if $s : I \to \mathbb{R}^+$ and $s^{(-1)} : \mathbb{R}^+ \to I$ are the continuous and non-decreasing functions defined by

$$s(x) = t(1-x), \quad s^{(-1)}(u) = 1 - t^{(-1)}(u), \tag{2.1.14}$$

then by (2.1.4),

$$S(x,y) = s^{(-1)}(s(x) + s(y)). \tag{2.1.15}$$

This correspondence yields the s-norm counterparts of the preceding results for t-norms. We list these in

Theorem 2.1.14 *Suppose that $S : I^2 \to I$ satisfies the hypotheses of Lemma 2.1.13 and, further, that S is Archimedean. Then:*

(a) S admits the representation (2.1.15), where s is a continuous and strictly increasing function from I to $\mathbb{R}^+$, with $s(0) = 0$, and $s^{(-1)}$ is the **pseudo-inverse** *of s, i.e.,*

$$s^{(-1)}(u) = \begin{cases} s^{-1}(u), \, 0 \le u \le s(1), \\ 1, \qquad s(1) \le u \le \infty. \end{cases} \tag{2.1.16}$$

(The functions s and $s^{(-1)}$ are, respectively, the inner and outer additive generators of S.)

(b) *S is commutative, hence an s-norm, i.e. in $\mathcal{S}_{Ar}$.*

(c) *Conversely, if $s : I \to \mathbb{R}^+$ is continuous and strictly increasing, with $s(0) = 0$, and if $s^{(-1)}$ is defined by (2.1.16), then the function S defined on I^2 by (2.1.15) is an element of $\mathcal{S}_{Ar}$.*

(d) *S is strict, i.e., in $\mathcal{S}_{St}$, if and only if $s(1) = \infty$, in which case*

$$S(x, y) = s^{-1}(s(x) + s(y)). \tag{2.1.17}$$

(e) *(I, S) is iseomorphic either to (I, P^*) or to (I, W^*), according as S is strict or not.*

(f) *$S = T^*$ if and only if S and T have additive generators related by (2.1.14).*

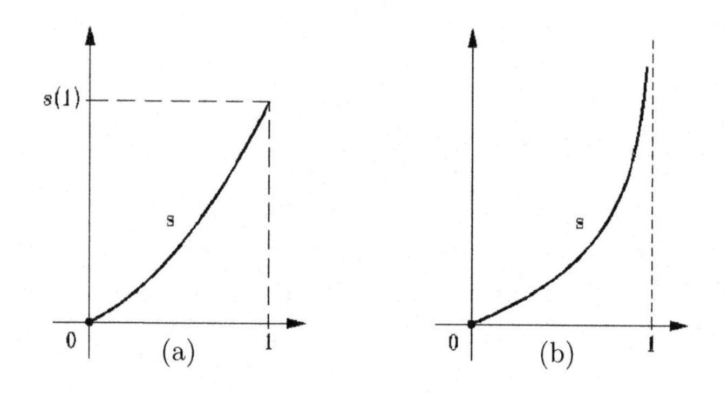

Figure 2.1.5. Inner additive generators of continuous Archimedean s-norms: (a) non-strict, (b) strict.

2.2 Additive and multiplicative generators

In this section we present some basic consequences of the representation theorem and relate properties of t-norms to properties of their generators.

Theorem 2.1.6 asserts that a two-place function T in $\mathcal{T}_{Ar}$ can be expressed as a composite of addition on $\mathbb{R}^+$ and the one-place functions t and $t^{(-1)}$. We next show that in (2.1.4) the operation of addition on $\mathbb{R}^+$ can be replaced by multiplication on I. This yields an alternate and frequently useful representation.

Suppose that T in $\mathcal{T}_{Ar}$ is additively generated by t. Define the functions h and $h^{(-1)}$ on I by

$$h(x) = e^{-t(x)}, \quad h^{(-1)}(x) = t^{(-1)}(-\log x). \tag{2.2.1}$$

The properties of t and $t^{(-1)}$ imply that $\operatorname{Ran} h \subseteq I$, that h is continuous and strictly increasing, that $h(1) = 1$, and that $h^{(-1)}$ is continuous and non-decreasing, with $h^{(-1)}(0) = 0$ and $h^{(-1)}(1) = 1$. Moreover, using (2.1.2) and (2.1.3), we find that for x in I,

$$h^{(-1)} \circ h = j_I \quad \text{and} \quad h \circ h^{(-1)}(x) = \max(h(0), x). \tag{2.2.2}$$

Now substituting $t(x) = -\log h(x)$ and $t^{(-1)}(u) = h^{(-1)}(e^{-u})$ into (2.1.4) yields

$$T(x, y) = h^{(-1)}(h(x)h(y)), \tag{2.2.3}$$

and thus we have

Theorem 2.2.1 *The function T belongs to $\mathcal{T}_{Ar}$ if and only if it admits the representation (2.2.3), where $h : I \to I$ is continuous and strictly increasing, with $h(1) = 1$, and $h^{(-1)}$ is given by*

$$h^{(-1)}(x) = \begin{cases} 0, & 0 \le x \le h(0), \\ h^{-1}(x), & h(0) \le x \le 1. \end{cases} \tag{2.2.4}$$

Moreover, T is strict if and only if $h(0) = 0$, in which case

$$T(x, y) = h^{-1}(h(x)h(y)). \tag{2.2.5}$$

The functions h and $h^{(-1)}$ in (2.2.3) are, respectively, **inner** and **outer multiplicative generators** of T.

Note that, for strict t-norms, Theorems 2.2.1 and 2.1.10 are identical.

Theorem 2.2.2 *Let T_1 and T_2 be elements of $\mathcal{T}_{Ar}$, with inner additive generators t_1 and t_2, respectively. Then:*

(a) $T_1 \le T_2$ if and only if the function $t_1 \circ t_2^{(-1)}$ is subadditive;
(b) $T_1 = T_2$ if and only if the restriction of $t_1 \circ t_2^{(-1)}$ to $[0, t_2(0)]$ is linear.

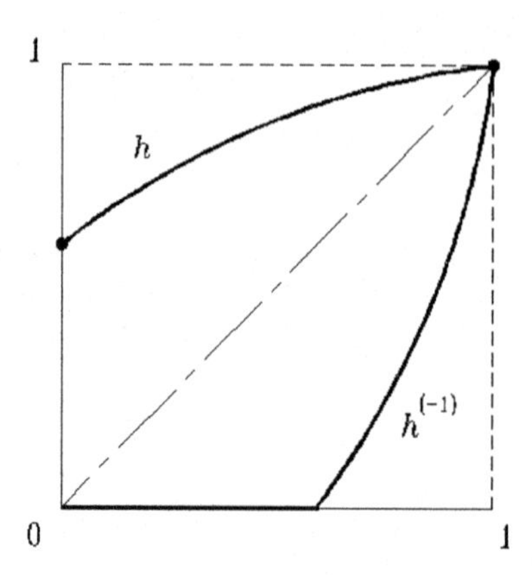

Figure 2.2.1. The multiplicative generators of a continuous Archimedean, non-strict t-norm.

Proof. Let $f = t_1 \circ t_2^{(-1)}$, and note that f is continuous and non-decreasing, with $f(0) = 0$. By (2.1.4), $T_1 \leq T_2$ if and only if

$$t_1^{(-1)}(t_1(x) + t_1(y)) \leq t_2^{(-1)}(t_2(x) + t_2(y)), \tag{2.2.6}$$

for all x, y in I. Setting $u = t_2(x)$ and $v = t_2(y)$, we have that (2.2.6) is equivalent to

$$t_1^{(-1)}(f(u) + f(v)) \leq t_2^{(-1)}(u + v), \tag{2.2.7}$$

for all u, v in $[0, t_2(0)]$. Observe that if $u > t_2(0)$ or $v > t_2(0)$, then each side of (2.2.7) is equal to 0. Suppose that $T_1 \leq T_2$. Then applying t_1 to both sides of (2.2.7) and using the fact that $t_1 \circ t_1^{(-1)}(w) \leq w$ for all w in $\mathbb{R}^+$, we get

$$f(u + v) \leq f(u) + f(v), \tag{2.2.8}$$

for all u, v in $\mathbb{R}^+$, i.e., f is subadditive. Conversely, if f satisfies (2.2.8), then applying $t_1^{(-1)}$ to both sides and noting that $t_1^{(-1)} \circ f = t_2^{(-1)}$ yields (2.2.7), and completes the proof of (a).

Next, replacing the inequality in (2.2.6) and (2.2.7) by equality, we have that $T_1 = T_2$ if and only if

$$f(u+v) = t_1 \circ t_1^{(-1)}(f(u) + f(v)), \tag{2.2.9}$$

for all u, v in $[0, t_2(0)]$. Suppose that the restriction of f to $[0, t_2(0)]$ is linear. Then since f is continuous, we must have $f(u) = (t_1(0)/t_2(0))u$ for u in $[0, t_2(0)]$, from which (2.2.9) follows readily, whence $T_1 = T_2$. Conversely, suppose that $T_1 = T_2$. Let $g : [0, t_1(0)] \to [0, t_2(0)]$ be the inverse of the restriction of f to $[0, t_2(0)]$. Then, letting $w = f(u)$ and $z = f(v)$, (2.2.9) yields

$$g \circ t_1 \circ t_1^{(-1)}(w + z) = g(w) + g(z),$$

for all w, z in $[0, t_1(0)]$, whence

$$g(w+z) = g(w) + g(z), \qquad \text{whenever } 0 \le w + z \le t_1(0). \tag{2.2.10}$$

Since the only continuous solutions of this conditional Cauchy equation are linear [Aczél (1966)], so is the restriction of f. □

It is frequently desirable to compare t-norms by way of their generators. Part (a) of Theorem 2.2.2 is of limited value in this regard because the subadditivity of $t_1 \circ t_2^{(-1)}$ is generally difficult to verify directly. However, several useful tests for subadditivity in this context have been developed. The first of these is a consequence of the well-known result that a function f on $\mathbb{R}^+$ is subadditive if it is concave and such that $f(0) = 0$ [Hille and Phillips (1957), Theorem 7.2.5].

Corollary 2.2.3 *Under the hypotheses of Theorem 2.2.2, if $t_1 \circ t_2^{(-1)}$ is concave, then $T_1 \le T_2$.*

A second and highly efficient test, originally due to R. Cooper [1927] (see also [Genest and MacKay (1986b)]), is the following:

Corollary 2.2.4 *Under the hypotheses of Theorem 2.2.2, if t_1/t_2 is non-decreasing on $(0, 1)$, then $T_1 \le T_2$.*

Proof. Define $g : (0, \infty) \to (0, \infty)$ by $g(u) = t_1 \circ t_2^{(-1)}(u)/u$. Since $g \circ t_2 = t_1/t_2$ and t_2 is decreasing, it follows that g is non-increasing on

$(0, t_2(0))$ and hence on $(0, \infty)$. Therefore, for all $u, v > 0$,

$$u[g(u+v) - g(u)] + v[g(u+v) - g(v)] \leq 0,$$

or, equivalently,

$$(u+v)g(u+v) \leq ug(u) + vg(v),$$

which asserts that $t_1 \circ t_2^{(-1)}$ is subadditive, completing the proof. □

The converses of Corollaries 2.2.3 and 2.2.4 are false. For example, let φ be the function given by

$$\varphi(u) = \begin{cases} 1 - (1-u)^2, & 0 \leq u \leq 1, \\ 2 - (2-u)^2, & 1 \leq u \leq 2, \\ u, & 2 \leq u, \end{cases}$$

let T_2 be any strict t-norm, let t_2 be an inner additive generator of T_2, and let T_1 be the strict t-norm generated by $t_1 = \varphi \circ t_2$. It is easily checked that φ is subadditive. Since $\varphi = t_1 \circ t_2^{(-1)}$, it follows that $T_1 \leq T_2$, even though φ is not concave. Moreover, t_1/t_2 fails to be non-decreasing; in fact, for $t_2(x)$ in $[1, 2]$, $t_1(x)/t_2(x) = (\varphi \circ t_2(x))/t_2(x) = (-2/t_2(x)) + 4 - t_2(x)$, which is strictly decreasing when x is in $[t_2^{-1}(2), t_2^{-1}(1)]$.

A third test, weaker than either of the previous tests, but often easier to apply, is a slight extension of a result presented in [Genest and MacKay (1986b)] (see also page 106, item 148 of [Hardy, Littlewood and Pólya (1952)]).

Corollary 2.2.5 *Under the hypotheses of Theorem 2.2.2, if t_1 and t_2 are continuously differentiable on $(0, 1)$, and t_1'/t_2' is non-decreasing on $(0, 1)$, then $T_1 \leq T_2$.*

Proof. Note first that, since t_1 and t_2 are decreasing and positive on (0,1), both $t_1' < 0$ and $t_2' < 0$ on (0,1). Let $g = t_1/t_2$ and $f = t_1'/t_2'$. Then:

(i) f is continuous and non-decreasing on (0,1), so that $\lim_{x \nearrow 1} f(x)$ exists (finite or infinite).

(ii) Since $\lim_{x \nearrow 1} t_1(x) = \lim_{x \nearrow 1} t_2(x) = 0$, L'Hôpital's Rule yields that $\lim_{x \nearrow 1} f(x) = \lim_{x \nearrow 1} g(x)$.

(iii) $g' = \left(\frac{t_1}{t_2}\right)' = \left(\frac{t_1'}{t_2'} - \frac{t_1}{t_2}\right) \cdot \frac{t_2'}{t_2} = (f-g)\frac{t_2'}{t_2}.$

By Corollary 2.2.4, it suffices to show that $g' \geq 0$, or equivalently since $t_2'/t_2 < 0$, that

$$f(x) - g(x) \leq 0, \qquad \text{for } 0 < x < 1.$$

Suppose to the contrary that there exists an x_0 in (0,1) such that $f(x_0) - g(x_0) > 0$. Then, since f is non-decreasing,

$$g(x_0) < f(x_0) \leq \lim_{x \nearrow 1} f(x) = \lim_{x \nearrow 1} g(x).$$

Furthermore $g'(x_0) < 0$. Thus there exists an x_1 in $(x_0, 1)$ such that $g(x_1) < g(x_0)$ and $g'(x_1) = 0$. But then (iii) and

$$g(x_1) < g(x_0) < f(x_0) \leq f(x_1)$$

together yield $g'(x_1) < 0$, a contradiction. □

Part (b) of Theorem 2.2.2 yields the fact that additive and multiplicative generators are essentially unique. More precisely, we have:

Corollary 2.2.6 *Let T be an element of $\mathcal{T}_{Ar}$.*

(a) Both t_1 and t_2 are inner additive generators of T if and only if there is an $\alpha > 0$ such that

$$t_2(x) = \alpha t_1(x), \quad t_1^{(-1)}(u) = t_2^{(-1)}(\alpha u). \tag{2.2.11}$$

(b) Both h_1 and h_2 are inner multiplicative generators of T if and only if there exists an $\alpha > 0$ such that

$$h_2(x) = [h_1(x)]^\alpha, \quad h_1^{(-1)}(x) = h_2^{(-1)}(x^\alpha). \tag{2.2.12}$$

Proof. Part (a) follows immediately from Theorem 2.2.2(b), (2.1.2) and the fact that $t_2 \circ t_2^{(-1)}$ is increasing. Part (b) follows from part (a) and (2.2.1). □

It follows that every non-strict Archimedean t-norm T has an inner additive generator t satisfying $t(0) = 1$. This normalization yields Ran t = Dom $t^{-1} = I$ and is frequently useful, as, for instance, in the following

Theorem 2.2.7 *Let T_1 and T_2 be continuous Archimedean non-strict t-norms with inner additive generators t_1 and t_2, respectively, normalized so that $t_1(0) = t_2(0) = 1$. Then $\mathcal{Z}(T_1) = \mathcal{Z}(T_2)$ if and only if the function $\varphi = t_1 \circ t_2^{-1}$ satisfies the condition*

$$\varphi(u) + \varphi(1-u) = 1, \tag{2.2.13}$$

for all u in I, i.e., if and only if the graph of φ is symmetric with respect to the point $\left(\frac{1}{2}, \frac{1}{2}\right)$.

Proof. From (2.1.12) it follows that $\mathcal{Z}(T_1) = \mathcal{Z}(T_2)$ if and only if $t_1^{-1}(1 - t_1(x)) = t_2^{-1}(1 - t_2(x))$ for all x in I. Setting $u = t_2(x)$ and simplifying yields (2.2.13). □

The equality $\mathcal{Z}(T_1) = \mathcal{Z}(T_2)$ does not imply $T_1 = T_2$. (See Section 3.8 and Examples A.7.1 and A.7.2. of Appendix A.)

Lemma 2.2.8 *Let T in $\mathcal{T}_{Ar}$ and $\lambda \geq 1$ be given, and let t be an inner additive generator of T. Then T satisfies the Lipschitz condition*

$$T(y, z) - T(x, z) \leq \lambda(y - x), \tag{2.2.14}$$

whenever $x \leq y$ if and only if

$$t^{(-1)}(u + w) - t^{(-1)}(v + w) \leq \lambda[t^{(-1)}(u) - t^{(-1)}(v)], \tag{2.2.15}$$

whenever $0 \leq u \leq v$ and $w \geq 0$.

Proof. First of all, (2.2.15) always holds when any one of u, v or w is greater than $t(0)$. For if $v \geq u > t(0)$ or $w > t(0)$, then $t^{(-1)}(u + w) = t^{(-1)}(v + w) = 0$, while the right-hand side is non-negative; and if $v > t(0) \geq u$, then $t^{(-1)}(v + w) = t^{(-1)}(v) = 0$ and (2.2.15) is again immediate. Next, $u, v, w \leq t(0)$ and $u \leq v$ if and only if there exist x, y, z in I with $x \leq y$ such that $u = t(y)$, $v = t(x)$, $w = t(z)$. Hence (2.2.15) is equivalent to

$$t^{(-1)}(t(y) + t(z)) - t^{(-1)}(t(x) + t(z)) \leq \lambda[t^{(-1)}(t(y)) - t^{(-1)}(t(x))],$$

which in turn is equivalent to (2.2.14). □

Theorem 2.2.9 *A t-norm T in $\mathcal{T}_{Ar}$ is a copula if and only if it is additively generated by a convex function.*

Proof. According to Theorem 1.4.5, T is a copula if and only if it satisfies (2.2.14) with $\lambda = 1$. Thus, in view of Lemma 2.2.8, it suffices to show that $t^{(-1)}$ is convex if and only if

$$t^{(-1)}(v) + t^{(-1)}(u+w) \le t^{(-1)}(u) + t^{-1}(v+w), \tag{2.2.16}$$

for $u \le v$ and $w \ge 0$. (Note that any generator t is convex if and only if $t^{(-1)}$ is convex.) Suppose first that $t^{(-1)}$ satisfies (2.2.16), and choose any x, y such that $0 \le x < y$. Setting $u = x$, $v = (x+y)/2$, $w = (y-x)/2$ in (2.2.16), we obtain

$$2t^{(-1)}\left(\frac{x+y}{2}\right) \le t^{(-1)}(x) + t^{(-1)}(y),$$

whence $t^{(-1)}$, being continuous, is convex. Next, suppose $t^{(-1)}$ is convex. Fix any u, v, w such that $u \le v$ and $w \ge 0$, and let $\alpha = (v-u)/(v-u+w)$. Since $v = (1-\alpha)u + \alpha(v+w)$ and since $u + w = \alpha u + (1-\alpha)(v+w)$, we have

$$t^{(-1)}(v) \le (1-\alpha)t^{(-1)}(u) + \alpha t^{(-1)}(v+w),$$
$$t^{(-1)}(u+w) \le \alpha t^{(-1)}(u) + (1-\alpha)t^{(-1)}(v+w).$$

Adding these inequalities yields (2.2.16). □

This characterization of copulas is an old result [Schweizer and Sklar (1961)]; the extension to Lemma 2.2.8 is due to Y.-H. Shyu [1984]. Note that, in view of the t-norm boundary values, $\lambda = 1$ is the strongest possible such condition. Shyu also has obtained the following connection between (2.2.15) and the derivative of a generator, a result which we state without proof.

Theorem 2.2.10 *Let T belong to $\mathcal{T}_{Ar}$ and let t be an inner additive generator of T. Suppose that the derivative of t exists and is negative on $(0,1)$. Then T satisfies (2.2.14) if and only if $t'(y) \ge \lambda t'(x)$ whenever $x < y < 1$.*

The probabilistic interpretation of copulas leads to many natural questions about their properties as joint distribution functions. In particular, it is often important to know when the measure induced by a copula is singular. Here, for Archimedean copulas, an answer is provided by the following theorem, whose proof employs various well-known differentiability properties of convex functions (see, e.g., [Roberts and Varberg (1973); Boas (1972); or Hardy, Littlewood and Pólya (1952)]).

Theorem 2.2.11 *Suppose that C is an Archimedean copula with inner additive generator t. Then:*

(a) For x_0 in $(0,1)$, the C-measure carried by the level curve $C(x,y) = x_0$ is given by

$$t(x_0)\left[\frac{1}{t'(x_0-)} - \frac{1}{t'(x_0+)}\right].$$

In particular, if $t'(x_0)$ exists – which is the case for all but at most countably infinite set of points – then this C-measure is 0.

(b) If C is not strict, then the C-measure carried by the level curve $C(x,y) = 0$, i.e., by the boundary curve of $\mathcal{Z}(C)$, is equal to $-t(0)/t'(0+)$, and thus equal to 0 whenever $t'(0+) = -\infty$.

Proof. Note first that the level curve $C(x,y) = x_0$ joins the points $(x_0,1)$ and $(1,x_0)$.

Suppose now that $x_0 > 0$. Let $w = t(x_0)$, let n be a fixed positive integer and consider the partition of $[x_0,1]$ which is induced by the regular partition $\{0, \frac{1}{n}w, ..., \frac{k}{n}w, ..., w\}$ of the interval $[0,w]$, i.e., the partition $\{x_0, x_1, ..., x_n = 1\}$, where

$$x_{n-k} = t^{(-1)}\left(\frac{k}{n}w\right), \quad k = 0,1,...,n.$$

Since $w < t(0)$, using (2.1.3), we have

$$\begin{aligned} C(x_j,x_k) &= t^{(-1)}(t(x_j) + t(x_k)) \\ &= t^{(-1)}\left(\frac{n-j}{n}w + \frac{n-k}{n}w\right) = t^{(-1)}\left(w + \frac{n-j-k}{n}w\right). \end{aligned}$$

In particular, $C(x_j, x_{n-j}) = t^{(-1)}(w) = x_0$.

Denote the rectangle $[x_{k-1},x_k] \times [x_{n-k},x_{n-k+1}]$ by R_k and let $S_n = \bigcup_{k=1}^{n} R_k$ (see Figure 2.2.2).

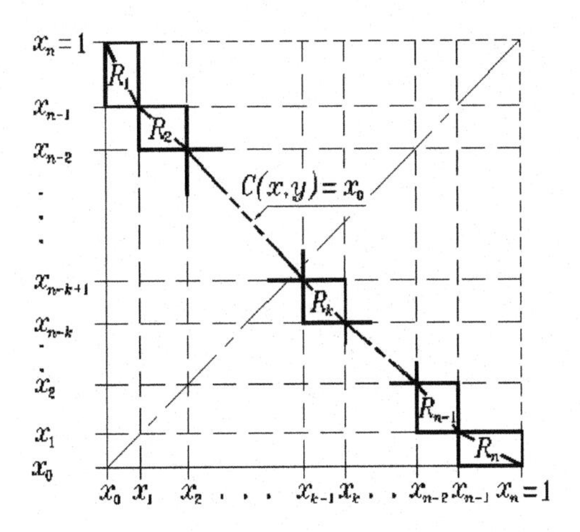

Figure 2.2.2. Typical approximation to $C(x, y) = x_0$.

From the convexity of $t^{(-1)}$ it follows that

$$0 \leq x_1 - x_0 \leq x_2 - x_1 \leq \cdots \leq x_n - x_{n-1} = 1 - x_{n-1},$$

and clearly, $\lim_{n\to\infty}(1 - x_{n-1}) = 1 - t^{(-1)}(0) = 0$. Hence the C-measure carried by the curve $C(x, y) = x_0$ is given by $\lim_{n\to\infty} m(S_n)$, where m denotes C-measure on I^2.

For each k we have,

$$\begin{aligned} m(R_k) &= C(x_{k-1}, x_{n-k}) - x_0 - x_0 + C(x_k, x_{n-k+1}) \\ &= \left[t^{(-1)}\left(w + \frac{w}{n}\right) - t^{(-1)}(w)\right] - \left[t^{(-1)}(w) - t^{(-1)}\left(w - \frac{w}{n}\right)\right]. \end{aligned}$$

Thus,

$$\begin{aligned} m(S_n) &= \sum_{k=1}^{n} m(R_k) \\ &= w\left[\frac{t^{(-1)}\left(w + \frac{w}{n}\right) - t^{(-1)}(w)}{w/n} - \frac{t^{(-1)}(w) - t^{(-1)}\left(w - \frac{w}{n}\right)}{w/n}\right], \end{aligned}$$

from which (a) follows on taking the limit as $n \to \infty$.

For a non-strict copula C and $x_0 = 0$, $t(0)$ is finite and $C(x,y) = 0$ on and below the level curve $C(x,y) = x_0$. Thus for each k, we have $m(R_k) = C(x_k, x_{n-k+1})$, from which, via the above argument, (b) readily follows. □

The special case (b) of Theorem 2.2.11 is due to C. Genest and R.J. MacKay [1986 a,b]. They obtained this result using a completely different argument, but under the stronger hypothesis that $t'' > 0$ on (0,1). In this case, the singular portion of C, if any, is concentrated on the boundary of $\mathcal{Z}(C)$. On the complement in I^2 of $\mathcal{Z}(C)$, C is absolutely continuous and possesses a density γ which is given by

$$\gamma(x,y) = -\frac{t''(C(x,y))t'(x)t'(y)}{[t'(C(x,y))]^3}.$$

For example, calculation of the measure $p = -t(0)/t'(0+)$ for the families of non-strict Archimedean copulas C_α listed in Table 2.6 yields: $p = 0$ for Families 1, 11, 14, 17, 24, and 25; $p = 1/\alpha$ for Families 2, 8, and 21; $p = -\alpha \log \alpha/(1-\alpha)$ for Family 7.

Another interesting example is provided by the strict copula C generated by

$$t(x) = \int_0^x \beta(z)dz - x + \frac{1}{2},$$

where β is the usual Cantor middle-thirds function on I. In this case $t'(x) = \beta(x) - 1$ is continuous, so the singular portion of C is concentrated on the boundary of $\mathcal{Z}(C)$ and has C-measure $-t'(0)/t(0) = \frac{1}{2}$, yet t'' fails to exist on the Cantor set.

Theorem 2.2.12 *Let C be an Archimedean copula with inner additive generator t. For x_0 in $[0,1)$, the C-measure carried by the region $\{(x,y)$ in $I^2 : C(x,y) \le x_0\}$ is given by*

$$x_0 - \frac{t(x_0)}{t'(x_0+)}.$$

We omit the proof, which is similar to that of Theorem 2.2.11 and is presented in [Nelsen (1999), Theorem 4.3.4].

We conclude this section with a discussion of the relationship between the differentiability properties of t-norms and their generators.

For any t-norm T and any point (x, y) in $(0,1)^2$, we denote the partial derivative of T with respect to its first place at (x, y) by $D_1T(x, y)$; similarly $D_2T(x, y)$ denotes the partial derivative of T with respect to its second place at (x, y). The corresponding two-place functions, wherever and whenever they are defined, will be denoted by D_1T and D_2T, respectively. The second order partial deriviatives of T at (x, y) are denoted by $D_{ij}T(x, y)$, $i, j = 1, 2$, and the corresponding functions by $D_{ij}T$, where $D_{12}T = D_2(D_1T)$; and similarly for higher order partial derivatives. Note that in view of the commutativity of T, $D_1T(x, y) = D_2T(y, x)$, $D_{11}T(x, y) = D_{22}T(y, x)$, and $D_{12}T(x, y) = D_{21}T(y, x)$. Thus, in particular, if $D_{12}T(x, y)$ exists for all (x, y) in $(0,1)^2$ then, automatically, $D_{12}T = D_{21}T$ on $(0,1)^2$.

Since T is non-decreasing in each place, it follows at once that, for each y in I, $D_1T(x, y)$ exists for almost all x in (0,1) and, similarly, that for each x in I, $D_2T(x, y)$ exists for almost all y in (0,1). Furthermore, since inner and outer additive generators are non-increasing, these functions are also differentiable almost everywhere on their respective domains. (Here "almost everywhere" refers to Lebesgue measure.)

Now suppose that T is a strict t-norm with inner additive generator t, so that

$$T(x, y) = t^{-1}(t(x) + t(y)), \text{ for all } (x, y) \text{ in } (0,1)^2. \tag{2.2.17}$$

Suppose further that $t'(x)$ exists and is different from 0 for all x in (0,1). Then D_1T and D_2T exist and are different from 0 on $(0,1)^2$. Specifically, letting $z = T(x, y)$, for all x, y in (0,1) we have

$$D_1T(x, y) = t'(x)/t'(z) \text{ and } D_2T(x, y) = t'(y)/t'(z). \tag{2.2.18}$$

If, in addition, t is twice differentiable on (0,1), then

$$\begin{aligned} D_{11}T(x, y) &= \frac{[t'(z)]^2 \cdot t''(x) - t''(z)[t'(x)]^2}{[t'(z)]^3}, \\ D_{12}T(x, y) &= -\frac{t'(x)t'(y)}{[t'(z)]^3} \cdot t''(z), \\ D_{22}T(x, y) &= \frac{[t'(z)]^2 \cdot t''(y) - t''(z)[t'(y)]^2}{[t'(z)]^3}. \end{aligned} \tag{2.2.19}$$

In general, if t is n-times differentiable, then all n-th order partial derivatives of T exist, i.e., the differentiability properties of t are inherited by T;

and if, in addition, $t'(x) \neq 0$ for all x in $(0,1)$, then expressions similar to (2.2.18) and (2.2.19) hold for these higher order partial derivatives as well. Note that if $t'' > 0$ on (0,1), then t' is increasing whence $t'(x) \neq 0$ for all x in (0,1).

In the other direction, since a well-behaved function can be the composite of ill-behaved functions (witness (2.5.5)), the matter is more involved. Here we have the following result, which is due to G. Szabó [1985]:

Theorem 2.2.13 *Let T be a strict t-norm with inner additive generator t and suppose that $D_1T(x,y)$ exists and is different from 0 for all (x,y) in $(0,1)^2$. Then $t'(x)$ exists and is different from 0 for all x in $(0,1)$.*

Proof. Fix a in (0,1). Then for each x in (0,1) distinct from a and each y in (0,1), we have

$$\frac{T(x,y)-T(a,y)}{x-a} = \frac{T(x,y)-T(a,y)}{t(x)-t(a)} \cdot \frac{t(x)-t(a)}{x-a},$$

from which, on noting that $t(x)-t(a) = t(T(x,y)) - t(T(a,y))$, we obtain

$$\frac{t(x)-t(a)}{x-a} = \frac{t(T(x,y))-t(T(a,y))}{T(x,y)-T(a,y)} \cdot \frac{T(x,y)-T(a,y)}{x-a}. \tag{2.2.20}$$

Since t is differentiable almost everywhere in (0,1), there is a point b in (0,1) such that $t'(T(a,b))$ exists. Letting $y=b$ in (2.2.20) and taking the limit as $x \to a$ therefore yields that $t'(a)$ exists and that

$$t'(a) = t'(T(a,b))D_1T(a,b).$$

Next, for any α in $(a,1)$ there is a β in (0,1) such that $T(\alpha,\beta) = T(a,b)$. Consequently,

$$t'(\alpha) = t'(T(\alpha,\beta))D_1T(\alpha,\beta) = t'(T(a,b))D_1T(\alpha,\beta).$$

Thus, if $t'(a) = 0$, then $t'(T(a,b)) = 0$, whence $t'(\alpha) = 0$ for all α in $(a,1)$, which is impossible since t is strictly decreasing. Thus $t'(a) \neq 0$ for all a in (0,1) and the theorem is proved. □

Corollary 2.2.14 *If $D_1T(x,y)$ exists and is different from 0 for all (x,y) in $(0,1)^2$, then the same is true of $D_2T(x,y)$. Moreover, the relations (2.2.18) hold, whence*

$$t'(x)/t'(y) = D_1T(x,y)/D_2T(x,y). \tag{2.2.21}$$

It is now a simple matter to pass from the existence of higher order partial derivatives of T to the existence of higher order derivatives of t. For letting, e.g., $y = 1/2$ in (2.2.21) yields

$$t'(x) = \frac{D_1T(x,1/2)}{D_2T(x,1/2)} \cdot t'(1/2),$$

from which everything follows.

In the case of a non-strict Archimedean t-norm, the situation is similar, but slightly more involved. Here the interior of I^2 divides into three parts: the interior of the zero-set $\mathcal{Z}(T)$ of T; the open region $(0,1)^2 \backslash \mathcal{Z}(T)$; and the boundary curve $t(x) + t(y) = t(0)$ (sans endpoints) which separates these regions. In the interior of $\mathcal{Z}(T)$, the partial derivatives of T clearly exist and are zero. For any (x, y) in the region $(0,1)^2 \backslash \mathcal{Z}(T)$, we have $t(x) + t(y) < t(0)$, so that $t^{(-1)}(t(x) + t(y)) = t^{-1}(t(x) + t(y))$. From this and the fact that, on the boundary curve, y is a strictly decreasing function of x, it readily follows that Theorem 2.2.12 and Corollary 2.2.13 remain valid for the restriction of T to $(0,1)^2 \backslash \mathcal{Z}(T)$. As regards D_1T and D_2T on the boundary curve of $\mathcal{Z}(T)$, it is clear that, if they exist, they must be 0. But then, on taking appropriate limits in (2.2.18), it follows at once that, if $t'(x)$ exists and is different from 0 for all x in (0,1), then $D_1T(x,y) = D_2T(x,y) = 0$ for all (x,y) on the boundary curve (sans endpoints) if and only if $\lim_{z \to 0+} t'(z) = -\infty$.

A version of Theorem 2.2.13 for a wider class of functions is given in [Jarai and Maksa (1985)].

All of the results of this section, including the construction of multiplicative generators, can be translated in the usual way to s-norms.

2.3 Extension to arbitrary closed intervals

In this section we consider semigroup operations defined on an arbitrary closed interval J of the extended reals – in particular, families of associative functions on J which parallel the t-norms and s-norms. We describe a natural way of identifying associative functions on I and J so that results concerning t-norms can be easily translated to J. Thus, the study of these families on any closed interval is reduced to the study of t-norms. The results are elementary, but needed in the sequel as well as in various applications, e.g., the generalized theory of information [Kampé de Fériet and Forte (1967); Kampé de Fériet, Forte and Benvenuti (1969); Kampé de Fériet (1974)] and the construction of products of metric spaces [Tardiff (1980)].

Let J be a closed subinterval of $\mathbb{R}$, and let ϕ be a continuous and strictly monotonic function from I onto J. For a given two-place function $H : J^2 \to J$, define the two-place function $B : I^2 \to I$ by

$$B(x, y) = \phi^{-1}[H(\phi(x), \phi(y))]. \tag{2.3.1}$$

We say that B is **generated from H by ϕ**. Since (2.3.1) is equivalent to

$$H(u, v) = \phi[B(\phi^{-1}(u), \phi^{-1}(v))], \tag{2.3.2}$$

it follows that H is generated from B by ϕ^{-1}.

The next lemma, whose verification is entirely elementary, shows that B inherits the algebraic, topological, and order properties of H, and vice versa.

Lemma 2.3.1 *Suppose that $B : I^2 \to I$ is generated from $H : J^2 \to J$ by ϕ, as in (2.3.1), where $\phi : I \to J$ is continuous, strictly monotonic, and onto. Then B is continuous, non-decreasing or strictly increasing in each place, commutative, or associative if and only if H possesses the corresponding property. Moreover, x is a null element, an identity, a nilpotent or an idempotent of the binary system (I, B) if and only if $\phi(x)$ is a null element, an identity, a nilpotent or an idempotent, respectively, of (J, H).*

Since ϕ is a homeomorphism between I and J and an algebraic isomorphism between (I, B) and (J, H), it is an iseomorphism which is either order-preserving or order-reversing.

Observe that the notion of generator as defined here generalizes the notions of the additive and multiplicative generators of strict t-norms and s-norms of the preceding sections. For if H is addition on $\mathbb{R}^+$, then ϕ is the inner additive generator of a strict t-norm or s-norm, according as ϕ is decreasing or increasing, respectively. An analogous statement holds for multiplicative generators when H is multiplication on I.

A more general way of generating new semigroups from old ones is presented in [Schweizer and Sklar (1983), (2005), §5.2], but will not be required here.

Our primary interest lies with those functions which generate, and are generated by, t-norms.

Definition 2.3.2 For $J = [a, b]$, $-\infty \leq a < b \leq \infty$, let $\mathcal{F}[a, b]$ and $\mathcal{G}[a, b]$

be the families of functions $F : J^2 \to J$ and $G : J^2 \to J$, respectively, which are non-decreasing in each place, commutative, and associative, and which satisfy the following boundary conditions:

$$F(u,a) = F(a,u) = a, \qquad F(u,b) = F(b,u) = u, \tag{2.3.3}$$

$$G(u,a) = G(a,u) = u, \qquad G(u,b) = G(b,u) = b, \tag{2.3.4}$$

for all u in J.

Note that $\mathcal{F}[0,1] = \mathcal{T}$ and $\mathcal{G}[0,1] = \mathcal{S}$, the set of t-norms and s-norms, respectively. We let $\mathcal{F}_{Co}[a,b]$ denote the set of continuous F in $\mathcal{F}[a,b]$, and $\mathcal{F}_{St}[a,b]$ the set of **strict** F, i.e., elements of $\mathcal{F}_{Co}[a,b]$ that are strictly increasing in each place on $(a,b]^2$. Similarly, we let $\mathcal{G}_{Co}[a,b]$ and $\mathcal{G}_{St}[a,b]$ denote the corresponding subsets of $\mathcal{G}[a,b]$. Thus, $\mathcal{F}_{Co}[0,1] = \mathcal{T}_{Co}$, etc.

Since the functions F and G are associative, F-powers and G-powers are well-defined, and therefore the Archimedean property makes sense: specifically, F in $\mathcal{F}[a,b]$ (resp., G in $\mathcal{G}[a,b]$) is **Archimedean** if, for every u, v in (a,b) there exists a positive integer m such that $u^m < v$ (resp., $u^m > v$). We let $\mathcal{F}_{Ar}[a,b]$ and $\mathcal{G}_{Ar}[a,b]$ denote the sets of Archimedean elements of $\mathcal{F}_{Co}[a,b]$ and $\mathcal{G}_{Co}[a,b]$, respectively.

Theorem 2.3.3 *Let $J = [a,b]$, $-\infty \le a < b \le \infty$; let $\phi : I \to J$ be continuous and strictly increasing, with $\phi(0) = a$ and $\phi(1) = b$; and let $\psi : I \to J$ be continuous and strictly decreasing, with $\psi(0) = b$ and $\psi(1) = a$.*

(a) Let $F, G : J^2 \to J$ be the functions generated from $T : I^2 \to I$ by ϕ^{-1}, ψ^{-1} respectively, i.e.,

$$F(u,v) = \phi[T(\phi^{-1}(u), \phi^{-1}(v))], \tag{2.3.5}$$

$$G(u,v) = \psi[T(\psi^{-1}(u), \psi^{-1}(v))]. \tag{2.3.6}$$

Then F belongs to $\mathcal{F}[a,b]$, $\mathcal{F}_{Co}[a,b]$, $\mathcal{F}_{Ar}[a,b]$, or $\mathcal{F}_{St}[a,b]$, and G belongs to $\mathcal{G}[a,b]$, $\mathcal{G}_{Co}[a,b]$, $\mathcal{G}_{Ar}[a,b]$, or $\mathcal{G}_{St}[a,b]$ if and only if T belongs to $\mathcal{T}$, $\mathcal{T}_{Co}$, $\mathcal{T}_{Ar}$, or $\mathcal{T}_{St}$, respectively.

(b) Part (a) remains valid upon replacing T by S, $\mathcal{T}$ by $\mathcal{S}$, and interchanging F and G, $\mathcal{F}$ and $\mathcal{G}$.

Proof. All of the statements are direct consequences of Lemma 2.3.1 and the fact that, in both cases, $[\phi(x)]^m = \phi(x^m)$ and $[\psi(x)]^m = \psi(x^m)$. □

Note that any one such ϕ or ψ yields a one-to-one correspondence between $\mathcal{T}$ and $\mathcal{F}[a,b]$ or $\mathcal{G}[a,b]$. In particular, when $J = I$ and $\psi = 1 - j$, (2.3.6) expresses the duality (1.3.12) between a t-norm T and its s-norm T^*. More generally, if F is generated from T by ϕ and if $\psi(x) = \phi(1-x)$, then F is generated from T^* by ψ.

Note also that on any interval J, $F = \text{Min}$ is generated from $T = \text{Min}$ by every ϕ and $G = \text{Max}$ from $T = \text{Min}$ by every ψ.

In the set $\mathcal{F}[-\infty, b]$ the analogues of t-norms having non-trivial zero sets are the functions that take on the value $-\infty$ at points in $(-\infty, b)^2$. We wish to avoid them since they are at best of minimal interest. Thus, we confine our attention to strict elements of $\mathcal{F}_{Ar}[-\infty, b]$. Similar comments apply to $\mathcal{G}[a, \infty]$.

The results in Section 1.3 admit immediate extensions to associative functions on any interval. For instance, it is easy to see that F belongs to $\mathcal{F}_{Ar}[a,b]$ if and only if its diagonal δ_F satisfies the inequality $\delta_F(u) < u$ for $a < u < b$, and G belongs to $\mathcal{G}_{Ar}[a,b]$ if and only if $\delta_G(u) > u$ for $a < u < b$. We now present the extension of the central Theorem 2.1.6.

Theorem 2.3.4 *(Representation Theorem)* *Let $J = [a,b]$, where $-\infty \leq a < b \leq \infty$. Suppose that $F : J^2 \to J$ is associative, is jointly continuous, satisfies the boundary conditions*

$$F(u,a) = F(a,u) = a, \quad F(b,b) = b,$$

and is Archimedean. Then:

(a) F admits the representation

$$F(u,v) = f^{(-1)}(f(u) + f(v)), \tag{2.3.7}$$

where f is a continuous and strictly decreasing function from J to $\mathbb{R}^+$, with $f(b) = 0$, and $f^{(-1)}$ is the **pseudo-inverse** *of f, i.e.,*

$$f^{(-1)}(r) = \begin{cases} f^{-1}(r), \ 0 \leq r \leq f(a), \\ a, \qquad f(a) \leq r \leq \infty. \end{cases} \tag{2.3.8}$$

Hence, F belongs to $\mathcal{F}_{Ar}[a,b]$.

(b) Conversely, if f and $f^{(-1)}$ are as above, then the function F defined on J^2 by (2.3.7) belongs to $\mathcal{F}_{Ar}[a,b]$.

(c) F belongs to $\mathcal{F}_{St}[a,b]$ if and only if $f(a) = \infty$, in which case $f^{(-1)} = f^{-1}$.

(d) *Similarly, $G : J^2 \to J$ is associative, jointly continuous, Archimedean, and satisfies*

$$G(a, a) = a, \qquad G(u, b) = G(b, u) = b,$$

if and only if G admits the representation

$$G(u, v) = g^{(-1)}(g(u) + g(v)), \tag{2.3.9}$$

where g is continuous and strictly increasing from J to $\mathbb{R}^+$, with $g(a) = 0$, and

$$g^{(-1)}(r) = \begin{cases} g^{-1}(r), & 0 \leq r \leq g(b), \\ b, & g(b) \leq r \leq \infty, \end{cases} \tag{2.3.10}$$

in which case G belongs to $\mathcal{G}_{Ar}[a, b]$. And G belongs to $\mathcal{G}_{St}[a, b]$ if and only if $g(b) = \infty$.

Proof. Suppose that F satisfies the hypotheses of the theorem. Choose $\phi : I \to J$ continuous, strictly increasing, and onto, and let T be given by $T(x, y) = \phi^{-1}[F(\phi(x), \phi(y))]$, i.e., let T be generated from F by ϕ. Then, by virtue of Lemma 2.3.1 and simple calculations, T satisfies the hypotheses of Theorem 2.1.6, whence T belongs to $\mathcal{T}_{Ar}$ and $T(x, y) = t^{(-1)}(t(x) + t(y))$. Let $f = t \circ \phi^{-1}$ and $f^{(-1)} = \phi \circ t^{(-1)}$. Then it is easy to check that f meets the stated requirements, that $f^{(-1)}$ satisfies (2.3.8), and that $f^{(-1)} = f^{-1}$ if and only if $t^{(-1)} = t^{-1}$. Moreover, we have

$$\begin{aligned} F(u, v) &= \phi[T(\phi^{-1}(u), \phi^{-1}(v))] \\ &= \phi \circ t^{(-1)}[t \circ \phi^{-1}(u) + t \circ \phi^{-1}(v)] \\ &= f^{(-1)}(f(u) + f(v)). \end{aligned}$$

The remainder of parts (a)-(c) then follow at once from Theorem 2.3.3. The proof part (d) is similar, by way of (2.3.6). □

In accordance with the special case $J = I$, we call f and $f^{(-1)}$ in (2.3.7) **inner** and **outer additive generators** of F, respectively, and similarly for g and $g^{(-1)}$ in (2.3.9). Note that inner additive generators of F and G are related to generators of their corresponding t-norms and s-norms via

$$f = t \circ \phi^{-1} = s \circ \psi^{-1} \tag{2.3.11}$$

and

$$g = s \circ \phi^{-1} = t \circ \psi^{-1}. \tag{2.3.12}$$

(The display (2.1.14) is a special case.)

The representation by multiplicative generators, Theorem 2.2.1, can be similarly extended from I to J, as can most of the remaining results in Section 2.2 – in particular, the fact that generators are essentially unique (Corollary 2.2.6). There is thus no real loss of generality in confining attention to I, and in the sequel we shall usually do so. The reader can use the results of this section to obtain the appropriate counterparts.

When J is compact, it is natural and often useful to identify semigroups on I and J by way of a linear mapping $\phi : I \to J$. Thus for $J = [a, b]$, $-\infty < a < b < \infty$, let

$$\phi(x) = a + (b - a)x.$$

Then (2.3.5) becomes

$$F(u, v) = a + (b - a)T\left(\frac{u - a}{b - a}, \frac{v - a}{b - a}\right), \tag{2.3.13}$$

or, equivalently,

$$T(x, y) = \frac{1}{b - a}[F(a + (b - a)x, a + (b - a)y) - a]. \tag{2.3.14}$$

Observe that the graph of F can be obtained from the graph of T by a dilation of $b - a$ followed by a translation.

If $T = \text{Min}$, then $F = \text{Min}$. Calculations of the functions F corresponding to the t-norms Z, W, and P yield:

$$Z_{[a,b]}(u, v) = \begin{cases} u, \ v = b, \\ v, \ u = b, \\ a, \text{ otherwise,} \end{cases} \tag{2.3.15}$$

$$W_{[a,b]}(u, v) = \max(u + v - b, a), \tag{2.3.16}$$

$$P_{[a,b]}(u, v) = a + \frac{(u - a)(v - a)}{b - a}. \tag{2.3.17}$$

For compact $[a, b]$, the duality between t-norms and s-norms expressed in (1.3.12) immediately extends to the duality between $\mathcal{F}[a, b]$ and $\mathcal{G}[a, b]$

via

$$F^*(u,v) = a + b - F(a + b - u, a + b - v). \tag{2.3.18}$$

To conclude this section, we present several important examples of continuous Archimedean F and G, together with their inner additive generators, on unbounded intervals.

(a) $J = [0, \infty]$:

$$G(u,v) = u + v, \quad g(u) = u;$$
$$F(u,v) = \frac{1}{\frac{1}{u} + \frac{1}{v}}, \quad f(u) = \frac{1}{u};$$
$$G(u,v) = u + v + uv, \quad g(u) = \log(1 + u).$$

(b) $J = [-\infty, 0]$:

$$F(u,v) = u + v, \quad f(u) = -u;$$
$$G(u,v) = \min[\log(e^u + e^v), 0], \quad g(u) = e^u.$$

(c) $J = [-\infty, \infty]$:

$$F(u,v) = u + v - \log(e^u + e^v), \quad f(u) = e^{-u};$$
$$G(u,v) = \log(e^u + e^v), \quad g(u) = e^u.$$

(d) $J = [1, \infty]$:

$$G(u,v) = uv, \quad g(u) = \log u.$$

2.4 Continuous, non-Archimedean t-norms

We begin by describing the so-called ordinal sum construction of t-norms (and s-norms) from other t-norms (and s-norms). This construction is the key that unlocks the structure of continuous, non-Archimedean t-norms: the principal result of this section characterizes elements of $\mathcal{T}_{Co}$ as ordinal sums of elements of $\mathcal{T}_{Ar}$.

Definition 2.4.1 Let $\{J_n\}$ be a countable collection of non-overlapping, closed, non-degenerate subintervals of I. With each $J_n = [a_n, b_n]$, associate a t-norm T_n and an s-norm S_n, and define the functions F_n and G_n on J_n^2 by

$$F_n(x,y) = a_n + (b_n - a_n)T_n\left(\frac{x-a_n}{b_n-a_n}, \frac{y-a_n}{b_n-a_n}\right), \tag{2.4.1}$$

$$G_n(x,y) = a_n + (b_n - a_n)S_n\left(\frac{x-a_n}{b_n-a_n}, \frac{y-a_n}{b_n-a_n}\right). \tag{2.4.2}$$

The **ordinal sum** of the (J_n, F_n) is the function $T : I^2 \to I$ given by

$$T(x,y) = \begin{cases} F_n(x,y), & (x,y) \text{ in } J_n^2,\ n = 1,2,\ldots, \\ \text{Min}(x,y), & \text{otherwise.} \end{cases} \tag{2.4.3}$$

The **ordinal sum** of the (J_n, G_n) is the function $S : I^2 \to I$ given by

$$S(x,y) = \begin{cases} G_n(x,y), & (x,y) \text{ in } J_n^2,\ n = 1,2,\ldots, \\ \text{Max}(x,y), & \text{otherwise.} \end{cases} \tag{2.4.4}$$

Theorem 2.4.2 *Let T be the ordinal sum (2.4.3). Then:*

(a) T is a t-norm, and the (J_n, F_n) are subsemigroups of (I, T).
(b) T belongs to $\mathcal{T}_{Co}$ if and only if each T_n belongs to $\mathcal{T}_{Co}$.
(c) T is a copula if and only if each T_n is a copula.
(d) If each T_n belongs to $\mathcal{T}_{Ar}$, then the idempotent set of T is $I \setminus \cup_n (a_n, b_n)$.
(e) Dual statements hold for the ordinal sum S in (2.4.4).

Proof. Since F_n is generated from T_n as in (2.3.13), all of the assertions in (a) and (b), apart from the associativity of T, are simple consequences of the definition of T and Theorem 2.3.3. Associativity follows from the comparison of $T(T(x,y),z)$ and $T(x,T(y,z))$ for the various cases according to the location of x, y, z relative to any one of the J_n and its complement. The copula inequality (1.4.1) follows from the consideration of the same cases, or from a probabilistic, measure-theoretic, argument. Part (d) is a consequence of Theorems 1.3.9 and 2.3.3, together with the fact that $\text{Min}(x,x) = x$. Finally, the proof of (e) is virtually identical. □

While (2.4.3) is the ordinal sum of the semigroups (J_n, F_n), it is convenient to refer to T, with slight abuse of language and notation, as "the ordinal sum of the (J_n, T_n)", or as "an ordinal sum of the t-norms T_n". Each T_n is called a **summand** of T.

Notice that the ordinal sum construction of Definition 2.4.1 includes two degenerate cases:

(1) If $\{J_n\}$ is empty, then $T = \text{Min}$ and $S = \text{Max}$.
(2) If $\{J_n\} = J_1 = I$, then $T = T_1$ and $S = S_1$.

To obtain the graph of the ordinal sum T in (2.4.3), begin with the graph of Min over I^2, and replace those portions over the squares J_n^2 by appropriately scaled graphs of the corresponding T_n.

As an elementary example, for $J_1 = \left[0, \frac{1}{4}\right]$, $J_2 = \left[\frac{1}{4}, \frac{1}{2}\right]$, and $J_3 = \left[\frac{3}{4}, 1\right]$, the ordinal sum of (J_1, W), (J_2, P), and (J_3, Z) is the t-norm

$$T(x,y) = \begin{cases} \text{Max}(x+y-1/4, 0), & (x,y) \text{ in } \left[0, \frac{1}{4}\right]^2, \\ 1/4 + (4x-1)(4y-1), & (x,y) \text{ in } \left[\frac{1}{4}, \frac{1}{2}\right]^2, \\ 3/4, & (x,y) \text{ in } \left(\frac{3}{4}, 1\right)^2, \\ \text{Min}(x,y), & \text{otherwise,} \end{cases}$$

which is depicted in Figure 2.4.1.

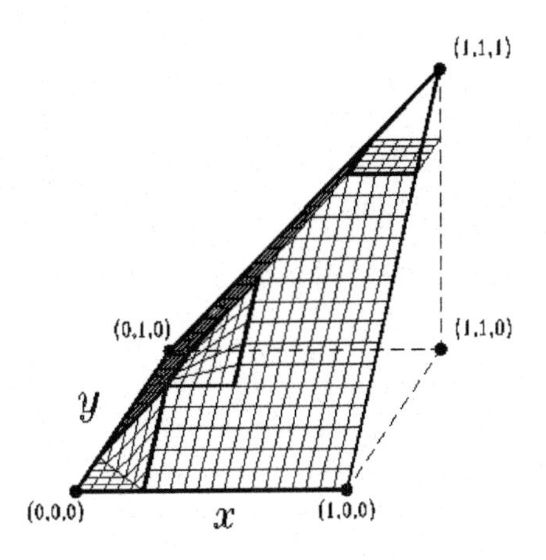

Figure 2.4.1. The graph of the ordinal sum T.

Note that if T is an ordinal sum and if $\mathcal{Z}(T)$, the zero set of T, is non-trivial, then it is a contraction of the zero set of the first component.

Within $\mathcal{T}_{Co}$, the converse of Theorem 2.4.2 is valid.

Theorem 2.4.3 *Suppose that $T: I^2 \to I$ satisfies the hypotheses of Lemma 2.1.1, i.e., T is jointly continuous, is associative, and satisfies the boundary conditions*

$$T(x,0) = T(0,x) = 0, \quad T(1,1) = 1.$$

Then T is expressible as an ordinal sum of elements of $\mathcal{T}_{Ar}$, whence T is in $\mathcal{T}_{Co}$. The dual statement for functions S which satisfy the hypotheses of Lemma 2.1.13 also holds.

Proof. Fix any T as above. By Lemma 2.1.1, $T(x,1) = T(1,x) = x$ and T is non-decreasing in each place. Let E_T denote the set of idempotents of T, i.e., $E_T = \{x : \delta_T(x) = x\}$. Since δ_T is continuous, E_T is closed. Hence $I \backslash E_T$ is a countable union of component intervals (a_n, b_n), $n = 1, 2, \ldots$. Let (a, b) be any one of these components, and let $J = [a, b]$. We first establish that $T(x,y) = \text{Min}(x,y)$ for

$$(i) \quad x \text{ in } E_T,\ y \text{ in } I \qquad \text{or} \qquad (ii) \quad x \text{ in } J,\ y \text{ in } I \backslash J.$$

In case (i), if $x \leq y$, then $x = T(x,x) \leq T(x,y) \leq T(x,1) = x$, so $T(x,y) = x$; if $x > y$, then by continuity $y = T(x,z)$ for some z, so $T(x,y) = T(x,T(x,z)) = T(T(x,x),z) = T(x,z) = y$. In case (ii), if $y < a$, then by (i) $y = T(a,y) \leq T(x,y) \leq y$, so $T(x,y) = y$; if $y > b$, then $x = T(x,b) \leq T(x,y) \leq x$, so $T(x,y) = x$. These arguments, with the roles of x and y reversed, yield that

$$T(x,y) = \text{Min}(x,y), \qquad \text{for } (x,y) \text{ in } I^2 \backslash \bigcup_n (a_n, b_n)^2. \tag{2.4.5}$$

Let F_n be the restriction of T to J_n^2. By (2.4.5), F_n satisfies the boundary conditions (2.3.3). Since F_n is non-decreasing, Ran $F_n = J_n$, so (J_n, F_n) is a subsemigroup of (I, T). Moreover, $F_n(x,x) < x$ for all x in (a_n, b_n). Hence by Theorem 2.3.4, F_n belongs to $\mathcal{F}_{Ar}(J_n)$. The function T_n defined by (2.4.1) therefore belongs to $\mathcal{T}_{Ar}$.

Finally, the dual result follows from (1.3.12) and the fact that if T is the ordinal sum of (J_n, T_n) then T^* is the ordinal sum of (J_n^*, T_n^*), where $J_n^* = [1 - b_n, 1 - a_n]$. □

This crucially important representation theorem is due to Mostert and Shields [1957]. (For special cases see [Climescu (1946)]; the first application to t-norms appears in [Schweizer and Sklar (1963)].)

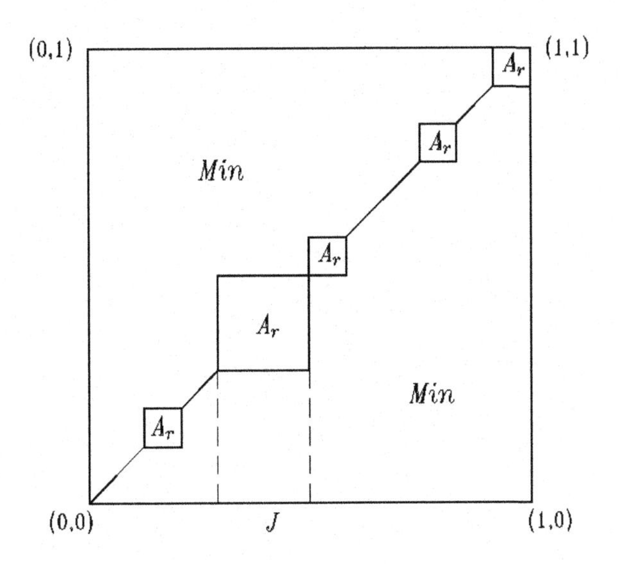

Figure 2.4.2. An ordinal sum.

Corollary 2.4.4 *If T satisfies the hypotheses of Lemma 2.1.1, then T is commutative (and dually for S).*

G. Krause (private communication) has shown that Corollary 2.4.4 can be established directly, i.e., without recourse to Theorems 2.1.6 and 2.4.3.

An extension of these representation theorems to certain discontinuous t-norms will be presented in the next section.

From Theorem 2.1.10 we obtain

Corollary 2.4.5 *Each T in $\mathcal{T}_{Co}$ is iseomorphic to an ordinal sum of t-norms, each of which is P or W (and dually for S in $\mathcal{S}_{Co}$, P^*, and W^*).*

Observe that the ordinal sum construction can be carried out on any interval $[a,b]$, and that, in view of the results of Section 2.3, Theorem 2.4.3 and its corollaries have their counterparts for $\mathcal{F}_{Co}[a,b]$ and $\mathcal{G}_{Co}[a,b]$.

We note further that every copula (associative or not) that has interior idempotents is representable as an "ordinal sum", i.e., $C(x,y) = \text{Min}(x,y)$ whenever (x,y) lies outside of the "diagonal squares" determined by its idempotents. To see this, suppose that $C(x_0,x_0) = x_0$, and choose (x,y) in $[0,x_0] \times [x_0,1]$. Since $C(x_0,y) = x_0$, it follows from (1.4.1) that $0 \le C(x_0,1) - C(x,1) - C(x_0,y) + C(x,y) = C(x,y) - x \le x - x = 0$, whence $C(x,y) = x = \mathrm{Min}(x,y)$. Similarly $C(x,y) = y$ for (x,y) in $[x_0,1] \times [0,x_0]$.

We conclude this section with two negative results. In 1957, V.I. Arnold showed that $\mathrm{Min}(x,y)$ cannot be represented as $f(g(x)+h(y))$ for any continuous functions f,g,h. More recently, G. Krause [1983] proved that, for a t-norm, the existence of even a single interior idempotent rules out such a representation and, a fortiori, a representation of the form (2.1.4). This last result admits the following slight improvement:

Theorem 2.4.6 *Suppose that $T : I^2 \to I$ satisfies the following boundary conditions:*

(i) $T(x,0) = 0$.
(ii) The functions $\phi(x) = T(x,1)$ and $\psi(x) = T(1,x)$ are one-to-one with $\phi \le j_I$ and $\psi \le j_I$.

If there exist continuous functions $g,h : I \to \mathbb{R}$ and $f : \mathbb{R} \to I$ such that

$$T(x,y) = f(g(x)+h(y)), \tag{2.4.6}$$

then $T(x,x) < x$ for $0 < x < 1$. Thus no element T of $\mathcal{T}_{Co} \backslash \mathcal{T}_{Ar}$ admits the representation (2.4.6).

Proof. Let J_1 and J_2 be the closed intervals $J_1 = g(I) + h(1)$ and $J_2 = g(1) + h(I)$. Since $\phi(x) = f(g(x)+h(1))$ is one-to-one, $h(1)$ must be finite and both g and $f|_{J_1}$ must be one-to-one, hence strictly monotonic. Since ϕ is continuous, one-to-one, and $\phi(0) = 0$, ϕ must be increasing. Hence, g and $f|_{J_1}$ must have the same direction. Similarly, h and $f|_{J_2}$ are strictly monotonic in the same direction. Note that therefore $g(1)+h(1)$ is a common endpoint of J_1 and J_2.

Suppose that g and h have opposite direction. Then choose the point $u_0 = g(0) + h(0)$ which lies between $g(0)+h(1)$ and $g(1)+h(0)$. Hence u_0 is in $J_1 \cup J_2$ and, by (i), $f(u_0) = 0$. It follows from the properties of f described in the preceding paragraph that either

$$T(0,1) = f(g(0)+h(1)) < f(u_0) = 0, \text{ for } u_0 \text{ in } J_1$$

or

$$T(1,0) = f(g(1)+h(0)) < f(u_0) = 0, \text{ for } u_0 \text{ in } J_2,$$

both of which are impossible.

Thus g and h have the same direction, and we may, without loss of generality, take g, h, and $f|_{J_1 \cup J_2}$ to be strictly increasing. Now fix an x in (0,1) and let $v = g(x) + h(x)$. If v is in $J_1 \cup J_2$, then

$$T(x,x) = f(v) < f(g(x) + h(1)) = T(x,1) \leq x;$$

if not, then (i) and the continuity of g guarantee the existence of a point x_0 in I for which $v = g(x_0) + h(0)$, in which case

$$T(x,x) = f(v) = f(g(x_0) + h(0)) = T(x_0, 0) = 0 < x. \qquad \square$$

The following examples show that neither of the boundary conditions (i) and (ii) in Theorem 2.4.6 can be eliminated.

Example 2.4.7 (a) For x in I let $g(x) = h(x) = \frac{1}{2}\log x$, and let $f(u) = e^u$. Then

$$T(x,y) = f(g(x) + h(y)) = \sqrt{xy}.$$

Now $T(x,0) = 0$, but $T(x,1) = T(1,x) = \sqrt{x} > x$ for $0 < x < 1$, and $T(x,x) = x$ for all x in I.

(b) For x in I let $g(x) = h(x) = 2x$, and let

$$f(u) = \begin{cases} u(2-u), & 0 \leq u \leq 2, \\ \frac{u}{2} - 1, & 2 \leq u \leq 4. \end{cases}$$

A short calculation yields

$$T(x,y) = f(g(x) + h(y)) = \max[x + y - 1, 4(x+y)(1-x-y)].$$

Now $T(x,1) = T(1,x) = x$, but $T(x,0) = 4x(1-x)$, and $T\left(\frac{7}{16}, \frac{7}{16}\right) = \frac{7}{16}$.

Finally we present the following strengthened version of Arnold's result concerning the non-representability of Min.

Theorem 2.4.8 *The t-norm Min does not admit a representation*

$$Min(x,y) = f(g(x) + h(y)) \tag{2.4.7}$$

for $g, h : I \to \mathbb{R}$ and $f : \mathbb{R} \to I$ satisfying either of the following sets of conditions:

(i) Ran g and Ran h are intervals;
(ii) f, g, and h are monotonic.

Proof. Observe that g and h are necessarily one-to-one; for if $g(x) = g(y)$, then $x = \mathrm{Min}(x, 1) = f(g(x)+h(1)) = f(g(y)+g(1)) = \mathrm{Min}(y, 1) = y$, and similarly for h.

In case (i), choose x, y, z in (0,1) so that $g(y)-g(0) = h(x)-h(z)$. Then (2.4.7) yields the contradiction

$$0 < \mathrm{Min}(y, z) = f(g(y) + h(z)) = f(g(0) + h(x)) = \mathrm{Min}(0, x) = 0.$$

In case (ii), since Min is non-decreasing in each place, it follows that the direction of monotonicity of f, g and h must be the same. Suppose, without loss of generality, that they are non-increasing. Then g and h are strictly decreasing. Let x in $(0, 1)$ be a point of continuity of h, choose y in $(x, 1]$, then choose $z < x$ so that $h(z) - h(x) < g(x) - g(y)$. This yields the contradiction

$$z = \mathrm{Min}(y, z) = f(g(y) + h(z)) \geq f(g(x) + h(x)) = x. \qquad \square$$

The argument in case (ii) can easily be refined to yield more: if g and h are monotonic, then f cannot be monotonic in any interval that contains two points of Ran g + Ran h.

2.5 Non-continuous t-norms

The primary focus of this book is on the set $\mathcal{T}_{Co}$ of continuous t-norms and, in view of the ordinal sum representation Theorem 2.4.3, on the set of continuous Archimedean t-norms $\mathcal{T}_{Ar}$. However, other t-norms merit study for a variety of reasons, quite apart from their intrinsic interest. First of all, in many applications the assumption of continuity is unwarranted. Second, discontinuous t-norms serve to illuminate the structure of $\mathcal{T}_{Co}$ by providing a source of examples and counterexamples of various phenomena. Third, certain discontinuous t-norms -notably Z- arise as limiting cases of families of elements of $\mathcal{T}_{Co}$.

The results of this section will be stated and proved for t-norms only; the dual results for s-norms, and the extensions to associative functions on arbitrary closed intervals, are straightforward. We begin with an extension of Theorem 1.3.9.

Theorem 2.5.1 *Let T be a t-norm. If*

$$\delta_T(x+) < x, \qquad \text{for} \quad 0 < x < 1, \tag{2.5.1}$$

then T is Archimedean. Conversely, if T is Archimedean, then T satisfies

(1.3.21), i.e.,

$$\delta_T(x) < x, \qquad \textit{for } 0 < x < 1; \tag{2.5.2}$$

and if there is an x_0 such that $\delta_T(x_0+) = x_0 > 0$, then $\delta_T(x) = x_0$ for some $x > x_0$.

Proof. Suppose that T satisfies (2.5.1) and consider x_0 in (0,1). Let $x_n = \delta_T^n(x_0)$ and $L = \lim x_n$. If $x_n > 0$, then $x_{n+1} = \delta_T(x_n) \leq \delta_T(x_n+) < x_n$. Thus, if $L > 0$, we have that $L < \delta_T(x_n) < x_n$ for all n, which implies that $\delta(L+) = L$, contrary to hypothesis. Hence $\lim_{n\to\infty} \delta_T^n(x_0) = 0$. Since x_0 is arbitrary, T is Archimedean by Lemma 1.3.8.

As for the converse, (2.5.2) is clearly necessary by Lemma 1.3.8, for if $\delta_T(x_0) = x_0$ for some x_0 then $\delta_T^n(x_0) = x_0$ for all n. Finally, suppose that $\delta_T(x_0+) = x_0 > 0$ but $\delta_T(x) > x_0 > 0$ for all $x > x_0$. Then for any $x_1 > x_0$, $\delta_T^n(x_1) > x_0$ for all n, and so $\lim_{n\to\infty} \delta_T^n(x_1) \geq x_0 > 0$, whence T is not Archimedean. □

Examples A.3.1 and A.3.2 of Appendix A show that Theorem 2.5.1 cannot be strengthened. Both t-norms satisfy (2.5.2) but not (2.5.1). Example A.3.1 is non-Archimedean; Example A.3.2 is Archimedean, yet $\delta_T\left(\frac{1}{2}+\right) = \frac{1}{2}$.

For continuous t-norms, strict monotonicity implies the Archimedean property (see Theorem 1.3.9). This is false in general.

Example 2.5.2 Let F be the contraction of P to $[1/2, 1]^2$ given by (2.4.1), so that

$$F(u, v) = uv + (1 - u)(1 - v), \text{ for } (u, v) \text{ in } \left[\frac{1}{2}, 1\right]^2.$$

Define T on $(0, 1]^2$ by

$$T(x, y) = \frac{1}{2^{p+q}} F(2^p x, 2^q y), \quad \frac{1}{2^{p+1}} < x \leq \frac{1}{2^p}, \quad \frac{1}{2^{q+1}} < y \leq \frac{1}{2^q},$$

for $p, q = 0, 1, 2, \ldots$.

It is easy to check that T is strictly increasing in each place on $(0, 1]^2$ and that T is discontinuous at the points in $(0, 1)^2$ which are on the lines $x = \frac{1}{2^n}$, $y = \frac{1}{2^n}$, $n = 1, 2, \ldots$. The associativity of T follows from that of F because, with x, y as above and $\frac{1}{2^{r+1}} < z \leq \frac{1}{2^r}$, we have

$T(T(x,y),z) = \frac{1}{2^{p+q+r}}F(F(2^p x, 2^q y), 2^r z)$, etc. However, T is not Archimedean since $\delta_T\left(\frac{1}{2}+\right) = \delta_F\left(\frac{1}{2}+\right) = \frac{1}{2}$, but $\delta_T(x) > \frac{1}{2}$ for all x in $\left(\frac{1}{2}, 1\right)$, contrary to Theorem 2.5.1.

Figure 2.5.1 exhibits a Venn diagram of the set $\mathcal{T}$ of t-norms, illustrating the relationships among the four subsets corresponding to the following conditions: continuity, strict monotonicity, the Archimedean property, and the lack of interior idempotents. Examples of t-norms belonging to each of the eight regions numbered on the diagram are given below.

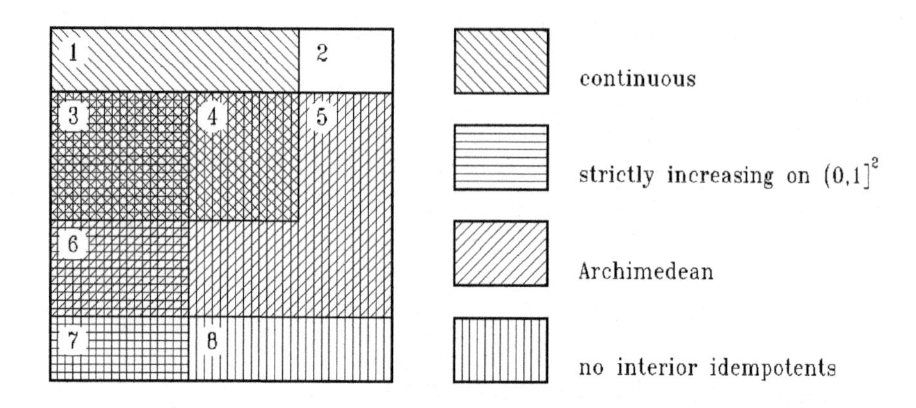

Figure 2.5.1. Venn diagram of $\mathcal{T}$.

1. Min 2. Ordinal sum of $\left(\left[0, \frac{1}{2}\right], Z\right)$ 3. P 4. W 5. Z 6. Example A.3.4 7. Example 2.5.2 8. Example A.3.1

We now address the question: to what extent do discontinuous t-norms admit, or not admit, representations of the kinds obtained in Sections 2.1 and 2.4? A more general version of Theorem 2.4.3 is given in

Theorem 2.5.3 *Let T be a t-norm. Suppose that the idempotent set of T is closed and that, for each idempotent x of T, the section σ_x is continuous. Then T is expressible as an ordinal sum of t-norms, none of which has any interior idempotents.*

Proof. A careful reading of the first part of the proof of Theorem 2.4.3 reveals that the arguments leading to (2.4.5) are valid *verbatim* under the present hypotheses. Since T is a t-norm, the restriction F_n of T to J_n^2 satisfies the conditions in part (a) of Theorem 2.3.3, whence the function T_n defined by (2.4.1) is a t-norm. □

Note that the summands in Theorem 2.5.3 need not be Archimedean or continuous.

Certain discontinuous Archimedean t-norms admit the representation (2.1.4). Here we consider t-norms generated by functions that are discontinuous at $x = 1$. Let T_0 be the continuous Archimedean t-norm generated by t_0, i.e.,

$$T_0(x, y) = t_0^{(-1)}(t_0(x) + t_0(y)).$$

For $c > 0$, let t_c and $t_c^{(-1)}$ be given by

$$t_c(x) = \begin{cases} t_0(x) + c, & 0 \le x < 1, \\ 0, & x = 1, \end{cases}$$

and

$$t_c^{(-1)}(u) = \begin{cases} 1, & 0 \le u \le c, \\ t_0^{(-1)}(u - c), & c \le u \le \infty. \end{cases}$$

Now define $T_c : I^2 \to I$ by

$$T_c(x, y) = t_c^{(-1)}(t_c(x) + t_c(y)). \tag{2.5.3}$$

Then $t_c \circ T_c(1, 1) = 0$ and $t_c \circ T_c(x, y) = \max[t_c(x) + t_c(y), t_c(0)]$ otherwise. An argument similar to the proof of Theorem 2.1.8 yields that T_c is an Archimedean t-norm which is continuous on $[0, 1)^2$ and discontinuous elsewhere. Moreover, T_c is strictly increasing if and only if T_0 is also. Observe that for (x, y) in $[0, 1)^2$, (2.5.3) is equivalent to

$$T_c(x, y) = t_0^{(-1)}(t_0(x) + t_0(y) + c), \tag{2.5.4}$$

and that $T_c = Z$ whenever $c \ge t_0(0)$.

An extension of the Representation Theorem 2.1.6 which includes this class of discontinuous t-norms is presented in Section 2.7. Here we give two illustrations.

Example 2.5.4

(a) Let $t_0 = -\log$, and set $k = e^{-c}$. Then (2.5.4) yields, for $0 < k < 1$,

$$T_k(x,y) = \begin{cases} kxy, & (x,y) \text{ in } [0,1)^2, \\ \mathrm{Min}(x,y), & \text{otherwise.} \end{cases}$$

(b) Let $t_0 = 1 - j_I$, and set $\lambda = c + 1$. Then (2.5.4) yields, for $\lambda > 1$,

$$T_\lambda(x,y) = \begin{cases} \max(x+y-\lambda, 0), & (x,y) \text{ in } [0,1)^2, \\ \mathrm{Min}(x,y), & \text{otherwise.} \end{cases}$$

Note that $T_\lambda = Z$ if and only if $\lambda \geq 2$.

In contrast to the non-representability results of Section 2.4, the following construction of J.Z. Yao and C.-H. Ling [1964] yields the representation

$$T(x,y) = \varphi^*(\varphi(x) + \varphi(y)) \tag{2.5.5}$$

for an *arbitrary* t-norm T. Here $\varphi^*\varphi = j_I$, but φ is always highly discontinuous and non-monotonic.

Let $\mathcal{H}$ be a Hamel basis for $\mathbb{R}$ consisting of positive numbers. Let φ be a one-to-one correspondence from (0,1) to $\mathcal{H}$, and set $\varphi(0) = \infty$, $\varphi(1) = 0$. Since the elements of $\mathcal{H} + \mathcal{H}$ are linearly independent, we have that $\mathcal{H} + \mathcal{H}$ and $\mathcal{H}$ are disjoint and that for any u in $\mathcal{H} + \mathcal{H}$ there exist unique x and y in $\mathcal{H}$ such that $u = \varphi(x) + \varphi(y)$. From this it follows that the function $\varphi^* : \mathbb{R}^+ \to I$ given by

$$\varphi^*(u) = \begin{cases} \varphi^{-1}(u), & u \text{ in } \mathcal{H} \cup \{0, \infty\}, \\ T(x,y), & u = \varphi(x) + \varphi(y) \text{ in } \mathcal{H} + \mathcal{H}, \end{cases}$$

and arbitrarily at remaining points of $\mathbb{R}^+$, is well-defined. The validity of the representation (2.5.5) is now immediate.

In Section 1.3 it was pointed out that neither the supremum nor the infimum of two t-norms need be a t-norm. However, E. O. Thorp [1960] proved the following result which is useful in dealing with limits of families of t-norms, as in Section 2.6.

Theorem 2.5.5 *If $\{T_\alpha : \alpha \text{ in } A\}$ is a totally ordered collection of left-continuous t-norms, then* $\sup_{\alpha \text{ in } A} T_\alpha$ *is a left-continuous t-norm.*

Proof. Let $T = \sup_{\alpha \text{ in } A} T_\alpha$. We have only to show that T is left-continuous, i.e., that

$$T(x,y) = \sup\{T(u,v) : 0 < u < x, 0 < v < y\}$$

for all x, y in $(0,1]$, and that T is associative. Using the fact that t-norms are non-decreasing and commutative, the left-continuity of T is easily established. To prove that T is associative, first note that, in the presence of the boundary conditions (1.3.1), associativity and commutativity are equivalent to the invariance of $T(T(x,y),z)$ under any permutation of (x,y,z). Thus, let $T(T(x_1,y_1),z_1)$ be given, let (x_2,y_2,z_2) be a permutation of (x_1,y_1,z_1), and suppose $\epsilon > 0$. Since $\{T_\alpha\}$ is totally ordered, there is a β such that

$$0 \le T[T(x_i,y_i),z_i] - T_\beta[T(x_i,y_i),z_i] < \frac{\epsilon}{2}, \quad i = 1,2;$$

and, from the definition of T and the fact that T_β is left-continuous, there is a $T_\alpha \ge T_\beta$ such that both $T(x_1,y_1) - T_\alpha(x_1,y_1)$ and $T(x_2,y_2) - T_\alpha(x_2,y_2)$ are so small that

$$0 \le T_\beta[T(x_i,y_i),z_i] - T_\beta[T_\alpha(x_i,y_i),z_i] < \frac{\epsilon}{2}, \quad i = 1,2.$$

Since $T_\alpha \ge T_\beta$, it follows that

$$T_\beta[T_\alpha(x_i,y_i),z_i] - T_\alpha[T_\alpha(x_i,y_i),z_i] \le 0, \quad i = 1,2.$$

Combining the inequalities (remembering that $T = \sup T_\alpha$), we have

$$0 \le T[T(x_1,y_1),z_1] - T_\alpha[T_\alpha(x_1,y_1),z_1] < \epsilon,$$

and

$$0 \le T[T(x_2,y_2),z_2] - T_\alpha[T_\alpha(x_2,y_2),z_2] < \epsilon.$$

But $T_\alpha[T_\alpha(x_1,y_1),z_1] = T_\alpha[T_\alpha(x_2,y_2),z_2]$, whence

$$|T[T(x_1,y_1),z_1] - T[T(x_2,y_2),z_2]| < \epsilon,$$

and the conclusion follows. □

The next result is from [Kolesárová (1999)]; a proof may be found in [Klement, Mesiar and Pap (2000)].

Theorem 2.5.6 *If an Archimedean t-norm is left-continuous, then it is continuous.*

Lastly, a long-standing question (Problem 5.8.1 of [Schweizer and Sklar (1983), (2005)]) asked whether, among t-norms satisfying $\delta_T(x) < x$ for $0 < x < 1$, continuity of T and δ_T are equivalent. G. Krause showed that this is not the case by constructing an intriguing and intricate counterexample, the so-called "Devil's Terraces", which employs classical Cantor sets. Thus

Example 2.5.7 There exists a discontinuous Archimedean t-norm whose diagonal is continuous.

Details of the construction are presented in Appendix B of [Klement, Mesiar and Pap (2000)]. A characterization of the class of t-norms having continuous diagonals remains open.

2.6 Families of t-norms

In this section we collect examples and families of examples of t-norms and related functions, many of which play prominent roles in applications and in subsequent chapters.

We begin with continuous Archimedean t-norms – the primary emphasis of this book. Elements of $\mathcal{T}_{Ar}$ can be constructed at will from appropriate decreasing one-place functions by way of the representation (2.1.4). For instance, $t(x) = \frac{1}{x} - 1$ generates the important strict t-norm

$$T(x, y) = \frac{xy}{x + y - xy},$$

which we designate by $P/(\Sigma - P)$.

A number of one-parameter families of generators t_α and the corresponding families of t-norms T_α have appeared in the literature in connection with various applications or as solutions of important functional equations and inequalities. Here we present a list of twenty-five such families and exhibit their basic properties in tabular form. Motivation for many of these families and references to the literature will be supplied as they arise in the sequel.

The first three columns of Table 2.6 contain the families, their inner additive generators, and the ranges of the (real) parameter α. The rest of the table is devoted to information which can be derived from results presented in preceding sections of this chapter.

(a) **Strictness**. According to Theorem 2.1.9(b), T_α is strict if and only if $t_\alpha(0) = \infty$.

(b) **The copula property**. For these families, it is usually a straightforward matter to apply Theorem 2.2.9: T_α is a copula precisely when $t''_\alpha > 0$ on (0,1).

(c) **Direction**. A family $\{T_\alpha\}$ is **positively** (resp., **negatively**) **directed** if $T_\alpha < T_\beta$ (resp., $T_\alpha > T_\beta$) whenever $\alpha < \beta$. Thus, for a positively directed family, the graphs of the T_α rise as α increases. Whether the direction of a given family is positive or negative is usually a matter of convenience in parametrization; if $0 < \alpha < \infty$ or $-\infty < \alpha < \infty$, the family can be redirected upon replacing α by $1/\alpha$ or $-\alpha$, respectively.

All of the families listed in the table except Family (2.6.10) are directed. For many of these it is quite difficult to establish the directional inequality from the expressions for $T_\alpha(x, y)$; in most cases it is much easier to verify one of the conditions on generators given in Corollaries 2.2.3-2.2.5. This leads to a variety of elementary exercises, some of which are decidely non-routine.

No two members of the Family (2.6.10) are comparable. For instance, a simple calculation yields that $T_\alpha(x, y) < T_{2\alpha}(x, y)$ if and only if $x^\alpha + y^\alpha < 1$.

(d) **Limiting t-norms**. If γ is an end point of the parameter interval, we denote by T_γ the function

$$T_\gamma(x, y) = \lim_{\alpha \to \gamma} T_\alpha(x, y)$$

(or the appropriate one-sided limit). For directed families Theorem 2.5.5 ensures that the limit is a t-norm, but not necessarily continuous or Archimedean. The limiting t-norms in the table can be obtained by standard methods which include L'Hospital's Rule.

#	$T_\alpha(x,y)$	$t_\alpha(x)$	Parameter Range
1	$(\max[x^{-\alpha}+y^{-\alpha}-1,0])^{-1/\alpha}$	$\frac{x^{-\alpha}-1}{\alpha}$	$(-\infty,\infty)$ $\alpha\neq 0$
2	$\max\left(1-[(1-x)^\alpha+(1-y)^\alpha]^{1/\alpha},0\right)$	$(1-x)^\alpha$	$(0,\infty)$
3	$xy/[1-\alpha(1-x)(1-y)]$	$\log\frac{1-\alpha+\alpha x}{x}$	$(-\infty,1]$
4	$\exp\left(-[(-\log x)^\alpha+(-\log y)^\alpha]^{1/\alpha}\right)$	$(-\log x)^\alpha$	$(0,\infty)$
5	$-\frac{1}{\alpha}\log\left[1+\frac{(e^{-\alpha x}-1)(e^{-\alpha y}-1)}{e^{-\alpha}-1}\right]$	$-\log\frac{e^{-\alpha x}-1}{e^{-\alpha}-1}$	$(-\infty,\infty)$ $\alpha\neq 0$
6	$1-[(1-x)^\alpha+(1-y)^\alpha-(1-x)^\alpha(1-y)^\alpha]^{1/\alpha}$	$-\log[1-(1-x)^\alpha]$	$(0,\infty)$
7	$\max[\alpha(x+y-1)+(1-\alpha)xy,0]$	$(\alpha-1)\log(\alpha+x-\alpha x)$	$[0,\infty)$
8	$\frac{\max[\alpha^2xy-(1-x)(1-y),0]}{\alpha^2-(\alpha-1)^2(1-x)(1-y)}$	$\frac{1-x}{1+(\alpha-1)x}$	$(0,\infty)$
9	$xye^{-\frac{1}{\alpha}\log x\log y}$	$\log(1-\frac{1}{\alpha}\log x)$	$(0,\infty)$
10	$xy/[1+(1-x^\alpha)(1-y^\alpha)]^{1/\alpha}$	$\log(2x^{-\alpha}-1)$	$(0,\infty)$

Table 2.6. Families #(2.6.n), $1\leq n\leq 10$.

#	Strict	Copula	Direction	Limiting t-norms			Special Cases
				$T_{-\infty}$	T_0	T_∞	
1	$\alpha \geq 0$	$\alpha \geq -1$	+	Z	P	Min	$T_{-1} = W$ $T_1 = \frac{P}{\Sigma - P}$
2	no	$\alpha \geq 1$	+		Z	Min	$T_1 = W$
3	yes	$\alpha \geq -1$	+	Z			$T_0 = P$ $T_1 = \frac{P}{\Sigma - P}$
4	yes	$\alpha \geq 1$	+		Z	Min	$T_1 = P$
5	yes	yes	+	W	P	Min	
6	yes	$\alpha \geq 1$	+		Z	Min	$T_1 = P$
7	no	$\alpha \leq 1$	−		P	Z	$T_1 = W$
8	no	$\alpha \geq 1$	+		Z	$\frac{P}{\Sigma - P}$	$T_1 = W$
9	yes	$\alpha \geq 1$	+		Z	P	
10	yes	$\alpha \leq 1$	None		P	P	

#	$T_\alpha(x,y)$	$t_\alpha(x)$	Parameter Range
11	$(\max[x^\alpha y^\alpha - 2(1-x^\alpha)(1-y^\alpha), 0])^{1/\alpha}$	$\log(2-x^\alpha)$	$(0,\infty)$
12	$\left(1+\left[\left(\frac{1}{x}-1\right)^\alpha+\left(\frac{1}{y}-1\right)^\alpha\right]^{1/\alpha}\right)^{-1}$	$(\frac{1}{x}-1)^\alpha$	$(0,\infty)$
13	$\left(\max\left[x^\alpha y^\alpha - \{(1-x^{2\alpha})(1-y^{2\alpha})\}^{1/2}, 0\right]\right)^{1/\alpha}$	$\arccos x^\alpha$	$(0,\infty)$
14	$\left(\max\left[\frac{x^\alpha y^\alpha - 2(1-x^\alpha)(1-y^\alpha)}{1-(1-x^\alpha)(1-y^\alpha)}, 0\right]\right)^{1/\alpha}$	$\alpha \arctan(1-x^\alpha)$	$(-\infty,\infty)$ $\alpha \neq 0$
15	$\exp\left(1-[(1-\log x)^\alpha+(1-\log y)^\alpha-1]^{1/\alpha}\right)$	$(1-\log x)^\alpha - 1$	$(0,\infty)$
16	$\left(1+\left[\left(x^{-1/\alpha}-1\right)^\alpha+\left(y^{-1/\alpha}-1\right)^\alpha\right]^{1/\alpha}\right)^{-\alpha}$	$\left(x^{-1/\alpha}-1\right)^\alpha$	$(0,\infty)$
17	$\left[\max\left(1-\left[\left(1-x^{1/\alpha}\right)^\alpha+\left(1-y^{1/\alpha}\right)^\alpha\right]^{1/\alpha}, 0\right)\right]^\alpha$	$\left(1-x^{1/\alpha}\right)^\alpha$	$(0,\infty)$
18	$\frac{1}{2}\left[x+y-1-\alpha\left(\frac{1}{x}+\frac{1}{y}-1\right) + \left\{\left[x+y-1-\alpha\left(\frac{1}{x}+\frac{1}{y}-1\right)\right]^2+4\alpha\right\}^{\frac{1}{2}}\right]$	$(\frac{\alpha}{x}+1)(1-x)$	$[0,\infty)$
19	$1-\exp\left(-\{[-\log(1-x)]^{-\alpha}+[-\log(1-y)]^{-\alpha}\}^{-\frac{1}{\alpha}}\right)$	$[-\log(1-x)]^{-\alpha}$	$(0,\infty)$
20	$\left(1+\frac{[(1+x)^{-\alpha}-1][(1+y)^{-\alpha}-1]}{2^{-\alpha}-1}\right)^{-\frac{1}{\alpha}}-1$	$-\log\frac{(1+x)^{-\alpha}-1}{2^{-\alpha}-1}$	$(-\infty,\infty)$ $\alpha \neq 0$

Table 2.6. Families #(2.6.n), $11 \leq n \leq 20$.

#	Strict	Copula	Direction	Limiting t-norms			Special Cases
				$T_{-\infty}$	T_0	T_∞	
11	no	$\alpha \leq \frac{1}{2}$	$-$		P	Z	
12	yes	$\alpha \geq 1$	$+$		Z	Min	$T_1 = \frac{P}{\Sigma - P}$
13	no	no	$-$		(A)	Z	
14	no	$\frac{-1}{\sqrt{2}} \leq \alpha \leq \frac{1}{\sqrt{2}}$	$\alpha < 0 : +$ $\alpha > 0 : -$	Z	P	Z	$T_{-1}(x,y) = \frac{x+y-1}{1-2(1-x)(1-y)}$, if $x + y \geq 1$
15	yes	yes	$+$		(B)	Min	$T_1 = P$
16	yes	$\alpha \geq 1$	$+$		(C)	Min	$T_1 = \frac{P}{\Sigma - P}$
17	no	$\alpha \geq 1$	$+$		Z	Min	$T_1 = W$
18	$\alpha > 0$	yes	$+$		W	$\frac{P}{\Sigma - P}$	
19	yes	no	$+$		Z	Min	
20	yes	yes	$+$	(D)	(E)	Min	$T_{-1} = P$, $T_1(x,y) = \frac{2xy}{1+x+y-xy}$

(A) $xy\exp(-2\sqrt{\log x \log y})$
(B) $xy\exp(-\log x \log y)$
(C) $xy/(x+y)$ on $[0,1)^2$
(D) $\max[\frac{1}{2}(x+y-1) + \frac{1}{2}xy, 0]$
(E) $2^{\log_2(1+x)\log_2(1+y)} - 1$

#	$T_\alpha(x,y)$	$t_\alpha(x)$	Parameter Range
21	$\max(1+\alpha/\log\left[e^{\alpha/(x-1)}+e^{\alpha/(y-1)}\right],0)$	$e^{\alpha/(x-1)}$	$(0,\infty)$
22	$\alpha/\log(e^{\alpha/x}+e^{\alpha/y}-e^{\alpha})$	$e^{\alpha/x}-e^{\alpha}$	$(0,\infty)$
23	$\left[\log\left(e^{x^{-\alpha}}+e^{y^{-\alpha}}-e\right)\right]^{-\frac{1}{\alpha}}$	$e^{x^{-\alpha}}-e$	$(0,\infty)$
24	$1-\left(1-\left\{\max\left([1-(1-x)^\alpha]^{1/\alpha}+[1-(1-y)^\alpha]^{1/\alpha}-1,0\right)\right\}^\alpha\right)^{1/\alpha}$	$1-[1-(1-x)^\alpha]^{1/\alpha}$	$(0,\infty)$
25	$\left[1-(1-x^\alpha)\sqrt{1-(1-y^\alpha)^2}-(1-y^\alpha)\sqrt{1-(1-x^\alpha)^2}\right]^{1/\alpha}$, if $(1-x^\alpha)^2+(1-y^\alpha)^2\le 1$, and 0 otherwise	$\arcsin(1-x^\alpha)$	$(0,\infty)$

Table 2.6. Families #(2.6.n), $21 \le n \le 25$.

The following criterion for $\lim_{\alpha\to\gamma} T_\alpha = \mathrm{Min}$, introduced in [Genest and MacKay (1986b)], is of particular interest. We present an elementary proof.

Theorem 2.6.1 *Let $\{T_\alpha\}$ be a family of Archimedean copulas with differentiable additive generators t_α. Then $\lim_{\alpha\to\gamma} T_\alpha = Min$ if and only if*

$$\lim_{\alpha\to\gamma} \frac{t_\alpha(x)}{t'_\alpha(x)} = 0, \qquad \textit{for } x \textit{ in } (0,1).$$

Proof. Suppose that the displayed condition holds. Fix an arbitrary x in (0,1) and choose ϵ in $(0,x)$. Then $0 < -t_\alpha(x)/t'_\alpha(x) < \epsilon$ for α sufficiently close to γ (in case γ is finite) or for $|\alpha|$ sufficiently large (in case γ is infinite). Since $x - t_\alpha(x)/t'_\alpha(x)$ is the x-intercept of the tangent to the graph of t_α at the point $(x, t_\alpha(x))$, the convexity of t_α immediately yields that for these α

$$t_\alpha\left(x+\frac{t_\alpha(x)}{t'_\alpha(x)}\right) > 2t_\alpha(x),$$

#	Strict	Copula	Direction	Limiting t-norms $T_{-\infty}$	T_0	T_∞	Special Cases
21	no	$\alpha \geq 2$	+		Z	Min	
22	yes	yes	+		$\frac{P}{\Sigma - P}$	Min	
23	yes	yes	+		P	Min	
24	no	$\alpha \geq 1$	+		Z	Min	$T_1 = W$
25	no	$0 < \alpha \leq 1$	–		P	Z	

whence

$$T_\alpha(x, x) = t_\alpha^{-1}\left(2t_\alpha(x)\right) > x + \frac{t_\alpha(x)}{t'_\alpha(x)} > x - \epsilon.$$

Hence $\lim_{\alpha\to\gamma} T_\alpha(x, x) = x$, so $T_\gamma =$ Min. The converse of the theorem follows by reversing the argument. □

A similar criterion for establishing that $\lim_{\alpha\to\gamma} T_\alpha = W$ is that $\lim_{\alpha\to\gamma} t_\alpha(x)/t'_\alpha(y) = x - 1$, for all x, y in $(0, 1)$.

(e) **Generators from t-norms.** It is often easy to find a generator t of a given Archimedean t-norm T or a family $\{T_\alpha\}$ by way of Corollary 2.2.14. As an illustration, consider $T = P/(\Sigma - P)$. Then $D_1T(x, y) = y^2/(x + y - xy)^2$, so that $t'(x)/t'(y) = y^2/x^2$, whence $t'(x) = c/x^2$ and, since $t(1) = 0$, $t(x) = k(\frac{1}{x} - 1)$.

To each family $\{T_\alpha\}$ there corresponds a dual family $\{T^*_\alpha\}$ of s-norms. Their properties can be gleaned from the table, the duality, and Theorem 2.1.14. Note, in particular, that the direction of $\{T^*_\alpha\}$ is opposite to that of $\{T_\alpha\}$.

There are various ways of building new families of t-norms from old, thereby enlarging the table. For instance, from a given generator $t(x)$, form the generators $1/t(1-x) - 1/t(0)$ and $\log[1+t(x)]$. Moreover, one can easily construct two-parameter families $T_{\alpha,\beta}$ by introducing a convenient second parameter $\beta > 0$ via either

$$t_{\alpha,\beta}(x) = t_\alpha(x^\beta) \quad \text{or} \quad t_{\alpha,\beta}(x) = [t_\alpha(x)]^\beta,$$

as well as by other means. Such two-parameter families arise naturally in applications, e.g., in statistical problems involving the modeling of dependence by copulas [Genest and Rivest (1993); Joe (1993); Nelsen (1997), (1999)]. Here a two-parameter family of copulas introduced by H. Joe [1993] is of particular interest; for, by enlarging the parameter space of Joe's family, we obtain a two-parameter family of t-norms that includes a surprising number of the prominent one-parameter families of t-norms and copulas as special or limiting cases.

Example 2.6.2 For $-\infty < \alpha < \infty$, $\alpha \neq 0$, and $-1 < \beta < \infty$, $\beta \neq 0$, let

$$T_{\alpha,\beta}(x,y) = \frac{1}{\beta}\left[\left(1 + \frac{[(1+\beta x)^{-\alpha} - 1][(1+\beta y)^{-\alpha} - 1]}{(1+\beta)^{-\alpha} - 1}\right)^{-1/\alpha} - 1\right]$$

be the strict t-norm generated by

$$t_{\alpha,\beta}(x) = -\log\frac{(1+\beta x)^\alpha - 1}{(1+\beta)^\alpha - 1}.$$

Properties of this family are summarized below, as well as in Figure 2.6.1, where the numerical labels refer to the corresponding families in Table 2.6 and the arrows indicate direction.

(a) *Copulas.*
$T_{\alpha,\beta}$ is a copula if and only if (i) $\alpha \leq -1$ or (ii) $\alpha > -1$ and $\beta \geq (1+\alpha)^{-1/\alpha} - 1$. (When $\alpha \leq -1$ and $-1 \leq \beta < 0$, we obtain the family introduced by H. Joe.)

(b) *Direction.*
   (i) For each fixed $\beta > 0$, the one-parameter family $T_{\alpha,\beta}$ is positively directed on $-\infty < \alpha < \infty$.
   (ii) For $\beta < 0$, $T_{\alpha,\beta}$ is negatively directed on $-\infty < \alpha < 1$.
   (iii) For each fixed $\alpha \geq 1$, $T_{\alpha,\beta}$ is positively directed on $0 < \beta < \infty$.
   (iv) For $\alpha \geq -1$, $T_{\alpha,\beta}$ is positively directed on $-1 < \beta < 0$.
   (v) For $\alpha < -1$, $T_{\alpha,\beta}$ is negatively directed on $-1 < \beta < 0$.

(c) *Special cases.*

(i) $\alpha = -1$: $T_{-1,\beta} = P$.
(ii) $\alpha = 1$: $T_{1,\beta}$ is Family (2.6.3), replacing α by $\beta/(1+\beta)$.
(iii) $\beta = -1, \alpha < 0$: $T_{\alpha,-1}$ is Family (2.6.6), replacing α by $-\alpha$.
(iv) $\beta = 1$: $T_{\alpha,1}$ is Family (2.6.20).

(d) *Limits.*

(i) $\beta \to 0:$ $T_{\alpha,0} = P$.
(ii) $\beta \searrow -1, \alpha > 0$: $T_{\alpha,-1} = Z$.
(iii) $\beta \to \infty$, $\alpha < 0$: $T_{\alpha,\infty} = P$.
(iv) $\beta \to \infty$, $\alpha > 0$: $T_{\alpha,\infty}$ is Family (2.6.1).
(v) $\alpha \to -\infty$, $\beta < 0$: $T_{-\infty,\beta} = Min$.
(vi) $\alpha \to -\infty$, $\beta > 0$: $T_{-\infty,\beta}$ is Family (2.6.7), replacing α by $\frac{1}{1+\beta}$.
(vii) $\alpha \to \infty$, $\beta > 0$: $T_{\infty,\beta} = Min$.
(viii) $\alpha \to \infty$, $\beta < 0$: $T_{\infty,\beta}$ is Family (2.6.7), replacing α by $\frac{1}{1+\beta}$.
(ix) For $\alpha \neq 0$, $\lim_{\beta \searrow 0} T_{\frac{\alpha}{\beta},\beta}$ is Family (2.6.5).
(x) $\alpha \to 0$:

$$T_{0,\beta}(x,y) = \exp\left[\frac{\log(1+\beta x)\log(1+\beta y)}{\log(1+\beta)} - \log\beta\right] - \frac{1}{\beta},$$

generated by $t_{0,\beta}(x) = \log\log(1+\beta) - \log\log(1+\beta x)$.

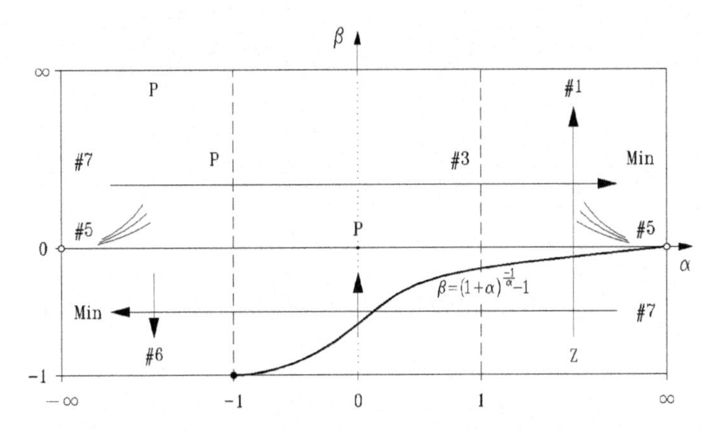

Figure 2.6.1. The parameter space for Example 2.6.2.

Next we present several families of discontinuous t-norms.

Example 2.6.3

(a) For $0 < \alpha < 1$, let
$$T_\alpha(x,y) = \begin{cases} \alpha, & (x,y) \text{ in } [\alpha,1)^2, \\ \operatorname{Min}(x,y), & \text{otherwise.} \end{cases}$$

(b) For $0 < \alpha < 1$, let
$$T_\alpha(x,y) = \begin{cases} \alpha, & (x,y) \text{ in } [\alpha,1)^2, \\ Z(x,y), & \text{otherwise.} \end{cases}$$

(c) For $0 < \alpha < \beta < 1$, let
$$T_{\alpha,\beta}(x,y) = \begin{cases} \alpha, & \alpha \le \operatorname{Min}(x,y) < \beta, \\ \operatorname{Min}(x,y), & \text{otherwise.} \end{cases}$$

(d) For $0 < \alpha < \beta < 1$, let
$$T_{\alpha,\beta}(x,y) = \begin{cases} \beta, & (x,y) \text{ in } [\beta,1)^2, \\ \operatorname{Min}(x,y), & \alpha < \operatorname{Min}(x,y) \le \beta, \\ Z(x,y), & \text{otherwise.} \end{cases}$$

(e) Let T be a given t-norm. For $0 < \alpha < 1$, let
$$T_\alpha(x,y) = \begin{cases} T(x,y), & T(x,y) > \alpha, \\ Z(x,y), & \text{otherwise.} \end{cases}$$

The methods of Section 2.3 allow us to construct families of examples in $\mathcal{F}[a,b]$ by way of $\mathcal{T}$. For example, one-parameter families $\{F_\alpha\}$ can be obtained in two ways by use of (2.3.5):

(i) from a family of t-norms T_α and a single generator ϕ;
(ii) from a single t-norm T and a family of generators ϕ_α.

For families in $\mathcal{F}_{Ar}[a,b]$, either route leads, by way of (2.3.11), to the generators $f_\alpha = t_\alpha \circ \phi^{-1} = t \circ \phi_\alpha^{-1}$. Similar remarks apply to $\mathcal{G}[a,b]$.

We conclude this section with a brief list of examples on $[0,\infty]$. Members of $\mathcal{F}[0,\infty]$ have been called **composition laws** [Kampé de Fériet (1974); Schweizer and Sklar (1983), (2005)], and are prominent in many applications. Some key families and their generators are the following:

(1) The generalized hyperbolic laws:
$$F_\alpha(u,v) = \frac{u^\alpha v^\alpha}{u^\alpha + v^\alpha}; \; f_\alpha(u) = \frac{1}{u^\alpha}, \; \alpha > 0; \qquad (2.6.26)$$

(2) The Wiener-Shannon laws of information theory:

$$F_\alpha(u,v) = \mathrm{Max}\left[-\frac{1}{\alpha}\log(e^{-\alpha u}+e^{-\alpha v}),0\right];$$

$$f_\alpha(u) = e^{-\alpha u},\ \alpha > 0; \tag{2.6.27}$$

(3)

$$\begin{aligned} F_\alpha(u,v) &= -\frac{1}{\alpha}\log\left[e^{-\alpha u}+e^{-\alpha v}-e^{-\alpha(u+v)}\right] \\ &= u+v-\frac{1}{\alpha}\log(e^{\alpha u}+e^{\alpha v}-1); \end{aligned}$$

$$f_\alpha(u) = -\log(1-e^{-\alpha u}),\ \alpha > 0. \tag{2.6.28}$$

Related families in $\mathcal{G}[0,\infty]$ are given by:

(4) The Minkowski laws:

$$G_\alpha(u,v) = (u^\alpha + v^\alpha)^{\frac{1}{\alpha}};\quad g_\alpha(u) = u^\alpha,\ \alpha > 0; \tag{2.6.29}$$

(5)

$$\begin{aligned} G_\alpha(u,v) &= \frac{1}{\alpha}\log\left(e^{\alpha u}+e^{\alpha v}-1\right); \\ g_\alpha(u) &= 1-e^{-\alpha u},\ \alpha > 0. \end{aligned} \tag{2.6.30}$$

2.7 Other representation theorems

In Sections 2.1-2.3 we derived the fundamental representation, Theorem 2.3.4, for the class of continuous, associative functions, defined on a closed interval J that have an identity at one endpoint of J (a null element at the other) and do not have idempotents in their interior. This result rests on the construction of outer generators of continuous Archimedean t-norms given in the proof of Theorem 2.1.6.

The purpose of this section is to describe some important modifications of this theorem and its proof. We present adaptations of the representation to other classes of associative functions and discuss a number of variants of the proof under weaker or alternate sets of initial assumptions. Although we give some indications, we omit proofs; but we do supply references to the rather extensive literature.

(A) Aczél's Representation Theorem

As pointed out in Section 1.1, in 1949 J. Aczél improved earlier versions of the representation theorem by weakening and/or eliminating various hypotheses. His result yields the representation for classes of associative functions that are continuous and strictly monotonic on an interval, without any assumptions whatsoever regarding boundary conditions or idempotents.

The monotonicity requirement is often expressed in algebraic language: for a given interval J, a function A is said to be **cancellative** (or **reducible**) on J^2 if for all x, y, z in J,

$$A(x,z) = A(y,z) \text{ or } A(z,x) = A(z,y) \text{ implies that } x = y.$$

In the presence of continuity in each place, this condition is equivalent to strict monotonicity in each place on J^2.

Theorem 2.7.1 *(Aczél's Representation Theorem [Aczél (1949), (1966)]) Let J be a proper, non-trivial subinterval of $\mathbb{R}$. Then the function $A : J^2 \to J$ is associative, continuous in each place, and cancellative on J^2 if and only if it admits the representation*

$$A(x,y) = a^{-1}(a(x) + a(y)), \tag{2.7.1}$$

where $a : J \to \mathbb{R}$ is continuous and strictly monotonic. Moreover, J must be open or half-open; and A admits a continuous and associative extension to $\overline{J}^2$ if and only if (2.7.1) holds for the continuous extension of a to $\overline{J}$.

Let $J = < c, d >$, $-\infty \le c < d \le \infty$, where $< c, d >$ denotes any one of the intervals (c, d), $[c, d)$, or $(c, d]$. Theorem 2.7.1 immediately implies that A is commutative and strictly increasing in each place on $< c, d >^2$. The assumption of strict monotonicity cannot be relaxed without imposing additional conditions; consider, for example, the function $A(x, y) = x$ on $< c, d >$. Clearly, a continuous t-norm or s-norm is in the above class if and only if it is strict.

Such a function A may or may not have an identity e; if it does, then e is clearly unique and may be an interior point or an endpoint of $[c, d]$. We examine these three situations in some detail. Note first that if e is an interior idempotent of A, then by (2.7.1), $a(e) = a(A(e, e)) = 2a(e)$, so $a(e) = 0$ and $A(x, e) = A(e, x) = a^{-1}(a(x)) = x$, whence e is the identity.

Interior identity. In this case (since $a(x^n) = na(x)$) Ran $a \supseteq (-\infty, \infty)$, and therefore, since a tends to $-\infty$ at one endpoint of $[c, d]$ and to $+\infty$ at the other, A cannot be continuously extended to all of $[c,d]^2$. We give two examples, each continuous on the indicated half-open squares:

(1) $A(x,y) = xy$ on $[0,\infty)^2$ or $(0,\infty]^2$. Here $e = 1$ and $a(x) = \log x$.
(2) $A(x,y) = xy/[xy+(1-x)(1-y)]$ on $[0,1)^2$ or $(0,1]^2$. Here $e = 1/2$ and $a(x) = \log(x^{-1} - 1)$.

Endpoint identity. If A is continuous on $[c,d]^2$ then either $a(c) = 0$ or $a(d) = 0$, whence Ran a is either $[0,\infty]$ or $[-\infty,0]$. This is precisely the strict case of Theorem 2.3.4.

No identity. If A is continuous on $[c,d]^2$ then there is no e in $[c,d]$ such that $a(e) = 0$, whence Ran a is either $[b,\infty]$ or $[-\infty,-b]$ for some $b > 0$, and without loss of generality we may set $b = 1$. Addition on $[1,\infty]$ is the obvious example. Two others are:

(1) For $p,q > 0$, $A(x,y) = (x^p+y^p+q)^{1/p}$ on $[0,\infty]$. Here $a(x) = \frac{x^p}{q}+1$.
(2) $A(x,y) = \frac{xy}{x+y}$ on I^2. Here $a(x) = 1/x$.

Observe that in all cases, no extension of A to $[c,d]^2$ can be strictly increasing on the entire boundary, because a takes an infinite value at one endpoint of $[c,d]$. In fact, that endpoint must be a null element of A.

The first case merits further comment. Associative functions on I whose identity is an interior point are necessarily discontinuous, in view of the following theorem due to R. J. Koch [1957] (see also page 169, item 17 of [Hofmann and Mostert (1966)]): *If a topological semigroup on a compact interval has an identity, it must be an endpoint.* Observe that this result justifies the original imposition of the t-norm and s-norm boundary conditions.

Now one can generate such functions $A : I^2 \to I$ from addition on $\mathbb{R}$ via (2.7.1), where $a : I \to \mathbb{R}$ is continuous and strictly decreasing, with $a(0) = \infty$ and $a(1) = -\infty$. (However, one must adopt a conventional value for $-\infty+\infty$ and $\infty+(-\infty)$.) Clearly, A is strictly increasing in each place on $(0,1)^2$ and is commutative. Since 0 is the identity of addition, it follows that $a^{-1}(0)$ is the identity of A. Note that the only discontinuities of A are the points (0,1) and (1,0).

It is easy to see that, with $\alpha = a^{-1}(0)$, the restrictions of A to $[0,\alpha]^2$ and $[\alpha,1]^2$ are elements of $\mathcal{F}_{St}[0,\alpha]$ and $\mathcal{G}_{St}[\alpha,1]$, respectively, and that Min $\le A \le$ Max on the remainder of I^2. (Contrast this with the situation in Corollary 2.7.11.)

For example, let

$$a(x)=\begin{cases}-\log(2x), & 0\le x\le \frac{1}{2},\\ \log(2-2x), & \frac{1}{2}\le x\le 1.\end{cases}$$

Then $A(x,y)$ is given in Figure 2.7.1; $\frac{1}{2}$ is the identity of A.

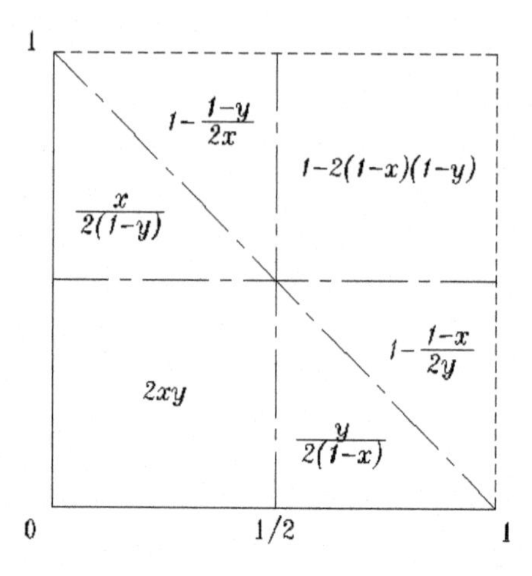

Figure 2.7.1. The function A.

The functions that satisfy the hypotheses of Aczél's Theorem and have an interior identity have been called **uninorms** and considered in [Fodor, Yager and Rybalov (1997)]. They are also discussed in Chapter 10 of [Klement, Mesiar and Pap (2000)], where further references to the literature may be found.

Generating associative functions in this way cannot be extended to the non-strict case. For if $a(0)>0$ and $a(1)<0$ are finite and if

$$a^{(-1)}(u)=\begin{cases}1, & -\infty\le u\le a(1),\\ a^{-1}(u), & a(1)\le u\le a(0),\\ 0, & a(0)\le u\le \infty,\end{cases}$$

then $B(x,y)=a^{(-1)}(a(x)+a(y))$ is continuous on I^2 and has identity in (0,1), and hence is not associative.

Aczél's proof of Theorem 2.7.1 [Aczél (1949)], which is based on the construction of the outer generator a^{-1}, is in the same spirit as the proof

Theorem 2.1.6 (and, in fact, was the motivation of this proof) but with strict monotonicity replacing the assumptions of an identity and the Archimedean property. The key idea is to define a^{-1} on rationals via $a^{-1}(m/n) = (z^{1/n})^m$ for a fixed non-idempotent z and then to extend a^{-1} by continuity. (Here $(\ )^m$ denotes the m-th A-power, as in the proof of Theorem 2.1.6, and $(\ )^{1/n}$ the corresponding inverse.) A complete and detailed exposition of this proof is given in [Aczél (1966)]; a somewhat simplified and streamlined version appears in [Aczél (1987)].

A novel and shorter proof of Theorem 2.7.1 is due to R. Craigen and Z. Páles [1989], who obtain the representation (2.7.1) by directly constructing the *inner* generator:

Theorem 2.7.2 *Let A be as in Theorem 2.7.1, and let z be a fixed non-idempotent element of A. Define*

$$a(x) = \inf\left\{\frac{k-m}{n} : k, m, n \text{ in } \mathbb{N} \text{ and } z^k < A(x^n, z^m)\right\},$$

in case $A(z,z) < z$, and reverse the inequality in case $A(z,z) > z$. Then a is a well-defined, continuous, strictly monotonic function on J that satisfies (2.7.1).

This construction cannot be adapted to the non-strict case.

(B) Strengthening Theorem 2.1.6

The representation (2.1.4) for continuous Archimedean t-norms can be obtained under hypotheses that are weaker than those given in Theorem 2.1.6. Indeed, one may strenghten the theorem by selecting various sets of initial assumptions and then either (1) deducing the remaining conditions directly, or (2) modifying the construction of an additive generator . We survey the most important results, all of which easily extend in the usual way to the classes $\mathcal{F}_{Ar}[a,b]$ and $\mathcal{G}_{Ar}[a,b]$.

The first result stems from a paper of Pedro Pi-Calleja [1954] on functional equations and the theory of magnitudes, in which the representation of certain associative functions on $\mathbb{R}^+$ (namely $\mathcal{G}_{st}(\mathbb{R}^+)$) was achieved by a clever construction of an inner additive generator . This construction was simplified, obtained under weaker continuity assumptions, and adapted to t-norms in [Alsina (1992)]. A further refinement of the argument yields:

Theorem 2.7.3 *Suppose that $T : I^2 \to I$ satisfies the following conditions:*

(i) T *is associative,*
(ii) T *is strictly increasing in each place on* $(0,1]^2$,
(iii) $T(x,1) = T(1,x) = x$, *for all* x *in* I,
(iv) δ_T *is continuous,*
(v) *For some* $\epsilon > 0$,

$$\lim_{y \nearrow 1} T(x,y) = x, \text{ for all } x \text{ in } [1-\epsilon, 1].$$

For any u *in* $(0,1]$ *and* v *in* $(0,1)$, *let* $N(u,v)$ *be the unique integer* $n \geq 0$ *for which* $v^{n+1} < u \leq v^n$; *and for* x *in* $(0,1]$, *let*

$$t(x) = \lim_{y \nearrow 1} \frac{N(x,y)}{N(z,y)},$$

where $0 < z < 1$ *is fixed. Then* t *is a well-defined, continuous, and strictly decreasing function on* $(0,1]$, *Ran* $t = [0, \infty)$, *and, with the convention that* $t(0) = \infty$, *we have*

$$T(x,y) = t^{-1}(t(x) + t(y)), \text{ for all } (x,y) \text{ in } I^2.$$

This result yields the strict case of Theorem 2.1.6 (Theorem 2.1.9(b)), but under the much weaker continuity assumptions (iv) and (v). The argument cannot, however, be adapted to the general Archimedan case, i.e., when (ii) is replaced by weak monotonicity.

Note the relation between Theorems 2.7.1 and 2.7.3: the continuity has been relaxed but the existence of an identity has been assumed. Finally, the construction of the inner additive generator is quite different from the corresponding construction in Theorem 2.7.2.

The next result, due to G. Krause [1981], extends the preceding theorem to Archimedean t-norms: the representation is derived from mild continuity and weak monotonicity assumptions. Moreover, one can even relax associativity to **power associativity**:

$$T(T(x^p, x^q), x^r) = T(x^p, T(x^q, x^r))$$

for all x in I and p, q, r in $\mathbb{N}$.

Theorem 2.7.4 *Suppose that* $T : I^2 \to I$ *satisfies the following conditions:*

(i) T *is power-associative,*
(ii) T *is non-decreasing in each place,*
(iii) $\delta_T(x) < x$ *for* $0 < x < 1$, *and* $\delta(1) = 1$,
(iv) δ_T *is continuous,*

(v) for some $\epsilon > 0$,

$$\lim_{y \nearrow 1} T(x,y) = x, \text{ for all } x \text{ in } [1-\epsilon, 1].$$

Then T admits the representation

$$T(x,y) = t^{(-1)}(t(x) + t(y)), \text{ for all } (x,y) \text{ in } [0,1)^2,$$

where t *and* $t^{(-1)}$ *are as given in Theorem 2.1.6. Moreover, if T has identity 1, then T belongs to* $\mathcal{T}_{Ar}$.

The proof of this theorem involves the construction of an outer generator, in the same spirit as the proof of Theorem 2.1.6, beginning with its values at dyadic rationals. The construction under the weaker hypotheses is, however, much more delicate. Note that, in view of Example 2.5.7, condition (v) cannot be eliminated.

Another derivation of the representation of power-associative functions, but beginning with continuity rather than monotonicity, was obtained by B. Bachelli [1986].

Theorem 2.7.5 *Suppose that* $T : I^2 \to I$ *satisfies the following conditions:*

(i) T is power-associative,
(ii) T is jointly continuous,
(iii) $T(x,1) = x$, *for all x in I,*
(iv) for all x in $(0,1)$ *and n in* $\mathbb{N}$, $x^{n+1} > 0$ *implies that* $x^{n+1} < x^n$.

Then T admits the representation (2.1.4), i.e., T belongs to $\mathcal{T}_{Ar}$.

Note that condition (iv) is stronger than the Archimedean property ($n = 1$). In fact, (iv) follows from the assumption that T is Archimedean and non-decreasing in one place.

Any inner additive generator t of a given continuous Archimedean t-norm T satisfies the Schröder equation $t(\delta_T(x)) = 2t(x)$ on a subinterval $[d,1]$ of I. We show in Section 3.8 that this equation always has other solutions which then generate different t-norms having diagonal δ_T. However, A. Sklar [1996b] has proved that, under certain natural conditions, t can be obtained as the solution of a pair of Schröder equations

$$\begin{aligned} t(f_2(x)) &= 2t(x), \\ t(f_3(x)) &= 3t(x), \end{aligned} \tag{2.7.2}$$

that arise in a natural way from an analysis of T-powers. For $n = 1, 2, \ldots$ let $f_n : I \to I$ be the T-power function $f_n(x) = x^n$ (in particular, $f_2 = \delta_T$). Assuming that f_{2^m} is invertible, let $f_{2^{-m}} = f_{2^m}^{-1}$. Let $D = \{2^m 3^n : m$ in $\mathbb{Z}$, n in $\mathbb{N}\}$, and for r in D, let $f_r = f_{2^m} \circ f_{3^n}$. Then we have

Theorem 2.7.6 *Suppose that $T : I^2 \to I$ satisfies the following conditions:*

(i) T is power-associative,
(ii) T is non-decreasing, or continuous, in each place,
(iii) $T(0,0) = 0$, $T(1,1) = 1$,
(iv) $T(x,y) < Min(x,y)$ for (x,y) in $(0,1)^2$;

and for the functions f_r and the set D defined in the preceding paragraph,

(v) f_2 is continuous, and there is a number d_2 in $(0,1)$ such that $f_2(x) = 0$ on $[0, d_2]$ and f_2 is strictly increasing on $[d_2, 1]$,
(vi) f_3 is continuous, and there is a number d_3 in $(d_2, 1]$ such that $f_3(x) = 0$ on $[0, d_3]$ and $f_3(x) > 0$ on $(d_3, 1]$,
(vii) for some x_0 in $(0,1)$, the set $\{f_r(x_0) : r$ in $D\}$ is dense in I.

Then T admits the representation (2.1.4), where t is a solution of the system (2.7.2).

In the original version of Theorem 2.1.6, Ling [1965] adopted the following hypotheses:

(1) T is associative,
(2) T is jointly continuous,
(3) $\delta_T(x) < x$, for $0 < x < 1$,
(4) $T(x,1) = x$, for all x in I,
(5) T is non-decreasing in each place.

It can be shown that if (4) and (5) are replaced by the weaker conditions:

(4′) $T(1,1) = 1$, $T(0,0) = 0$,
(5′) $T(x,0) \leq T(y,0)$, whenever $x \leq y$,

then these conditions, together with (3), yield the remaining boundary conditions in Lemma 2.1.1. Moreover, as was shown in [Bachelli (1986)], (5′) can be replaced by

(5″) for each $\epsilon > 0$ there is an x in $(0, \epsilon)$ such that $T(x,0) \neq x$.

Finally, we point out that characterizations of those continuous, monotonic functions that are representable in the form $g(f(x)+f(y))$, with continuous f and g, are presented in [Paganoni and Rusconi (1983); Rusconi (1992)].

(C) Some additional classes

We first consider the class of non-decreasing functions on I^2 for which there is no identity and for which 0 is the null element. If such a function admits the representation (2.7.1), then it has a generator a which is continuous and strictly decreasing from I onto $[1,\infty]$. Hence the function t given by $t(x) = a(x) - 1$ generates a strict t-norm; and since $a^{-1}(u) = t^{-1}(u-1)$, we have

$$A(x,y) = t^{-1}(t(x) + t(y) + 1).$$

The following theorem, due to A. Sklar (private communication), gives conditions under which t^{-1} can be replaced by $t^{(-1)}$.

Theorem 2.7.7 *Suppose that $B : I^2 \to I$ satisfies the following conditions:*

(i) B is associative,
(ii) B is non-decreasing in each place,
(iii) the function μ defined by

$$\mu(x) = \lim_{y \nearrow 1} B(x,y) = \lim_{y \nearrow 1} B(y,x)$$

is continuous, satisfies $\mu(x) < x$ for $0 < x \leq 1$, and is strictly increasing wherever it is positive,
(iv) δ_B is continuous, $\delta_B(x) < \mu(x)$ whenever $\mu(x) > 0$, and δ_B is strictly increasing on $[d,1]$, where $d = \sup\{x : \delta_B(x) = 0\}$, and $\delta_B^2(1) \geq d$,
(v) B is commutative on $[d,1]^2$,
(vi) $B(x,1) = B(1,x) = \mu(x)$, for all x in I.

Then B admits the representation

$$B(x,y) = t^{(-1)}(t(x) + t(y) + 1),$$

where t is the generator of a continuous Archimedean t-norm. Thus

$$B(x,y) = b^{(-1)}(b(x) + b(y)), \tag{2.7.3}$$

where $b(x) = t(x) + 1$ *for* $0 \leq x \leq 1$, *and* $b^{(-1)}(u) = t^{(-1)}(u-1)$ *for* $1 \leq u \leq \infty$.

This immediately yields the representation for a class of t-norms that are discontinuous (only) on the upper boundary. (See Example 2.5.4 and the discussion preceding it.) By redefining $b(1) = 0$, we obtain:

Corollary 2.7.8 *Suppose that* $B : I^2 \to I$ *satisfies conditions (i)-(v) of Theorem 2.7.7, and*

(vi) $B(x, 1) = B(1, x) = x$.

Then B *is a t-norm that is continuous on* $[0, 1)^2$ *(but not on the upper boundary), and* B *admits the representation (2.7.3), where* b *and* $b^{(-1)}$ *are as in Theorem 2.7.7, except that* $b(1) = 0$.

Discontinuous, associative functions B can also be generated from weakly monotonic functions b and their quasi-inverses. We present one such construction, which is a very special case of Theorem 5.2.1 in [Schweizer and Sklar (1983), (2005)].

Let $b : I \to \mathbb{R}^+$ be a continuous, *non-increasing* function with $b(1) = 0$, and let $b^* : [0, b(0)] \to I$ be a **quasi-inverse** of b, i.e., any function that satisfies $b(b^*(u)) = u$, for $0 \leq u \leq b(0)$. Note that b^* is strictly decreasing, is discontinuous at u_0 if and only if $b(x) = u_0$ on some subinterval of I, and thus is unique if and only if b is strictly decreasing. Now in case $b(0) < \infty$, extend b^* to $\mathbb{R}^+$ by defining $b^*(u) = 0$ for $b(0) < u \leq \infty$. Observe that $b(b^*(u)) = \min[u, b(0)]$ for all u in $\mathbb{R}^+$, and that $b^*(b(x)) = x$ if and only if x belongs to Ran b^*.

Theorem 2.7.9 *For any given* b *and* b^* *as described in the preceding paragraph, define the function* $B : I^2 \to I$ *via*

$$B(x, y) = b^*(b(x) + b(y)).$$

Then B *is associative, non-decreasing in each place, and commutative; and* B *has null element* 0. *Moreover,* B *is a t-norm if and only if* b *is strictly increasing.*

The reader is encouraged to consult the previously cited reference for the more general construction of semigroups. (Theorem 2.7.9 is the case $S = \mathbb{R}^+$, $T_1 = +$, $g = b$, $f = b^*$, and $a = 0$.)

To conclude this section, we consider a class of associative functions that may be viewed as a type of "ordinal sum" of a t-norm and an s-norm. We first characterize those associative functions on I^2 that satisfy boundary conditions weaker than the t-norm and s-norm conditions. This result is due to J.-L. Marichal [1998]; an alternate proof appears in [Sander (1998)].

Theorem 2.7.10 *The function $A : I^2 \to I$ is associative, continuous and non-decreasing in each place, and satisfies the boundary conditions*

$$A(0,0) = 0, \quad A(1,1) = 1$$

if and only if

$$(i)\ A(x,y) = \begin{cases} G(x,y), \text{ for } (x,y) \text{ in } [0,p]^2, \\ F(x,y), \text{ for } (x,y) \text{ in } [q,1]^2, \end{cases}$$

where $p = Min[A(0,1), A(1,0)]$, $q = Max[A(0,1), A(1,0)]$, G belongs to $\mathcal{G}_{C_0}[0,p]$, and F belongs to $\mathcal{F}_{C_0}[q,1]$,

and

(ii) for (x,y) in $I^2 \backslash ([0,p]^2 \cup [q,1]^2)$,

$$A(x,y) = \begin{cases} median[x, A(0,1), A(1,0)], \text{ if } A(0,1) \leq A(1,0), \\ median[y, A(0,1), A(1,0)], \text{ if } A(0,1) \geq A(1,0). \end{cases}$$

Proof. Straightforward calculations show that any such A satisfies all of the given conditions. We prove the converse for the case $p = A(0,1) \leq A(1,0) = q$, the argument in the other case being essentially identical.

The associativity of A yields that

$$A(0,p) = A(0, A(0,1)) = A(A(0,0),1) = A(0,1) = p.$$

Similarly

$$A(p,1) = p \text{ and } A(q,0) = A(1,q) = q.$$

Hence, since A is non-decreasing, we have

$$A(x,y) = \begin{cases} p, \text{ for } (x,y) \text{ in } [0,p] \times [p,1], \\ q, \text{ for } (x,y) \text{ in } [q,1] \times [0,q]. \end{cases}$$

Now consider any x in $[p, q]$. Since $A(p, 0) \leq A(p, 1) = p$ and $A(q, 0) = q$, the continuity of A implies that there exists a z in $[p, q]$ such that $A(z, 0) = x$, whence

$$A(x, 0) = A(A(z, 0), 0) = A(z, A(0, 0)) = A(z, 0) = x;$$

and similarly $A(x, 1) = x$. Monotonicity then yields that $A(x, y) = x$ for all (x, y) in $[p, q] \times I$. Therefore,

$$A(x, y) = \text{median}(x, p, q), \text{ for } (x, y) \text{ in } I^2 \backslash ([0, p]^2 \cup [q, 1]^2).$$

Let G and F be the restrictions of A to $[0, p]^2$ and $[q, 1]^2$, respectively. The preceding argument shows that G and F satisfy the boundary conditions

$$\begin{aligned} G(0, 0) = 0,\ G(x, p) = G(p, x) = p, \\ F(1, 1) = 1,\ F(x, q) = F(q, x) = q. \end{aligned}$$

Moreover, G and F must be associative, continuous, and non-decreasing. Hence, by (the extension of) Lemma 2.1.1 and its dual and Lemma 2.1.2, G and F belong to $\mathcal{G}_{C_0}[0, p]$ and $\mathcal{F}_{C_0}[q, 1]$, respectively. □

Theorem 2.7.10 admits an obvious extension to associative functions on any interval. Note that A is a t-norm or s-norm if and only if $p = q = 0$ or $p = q = 1$, respectively. It is important to observe that A is commutative

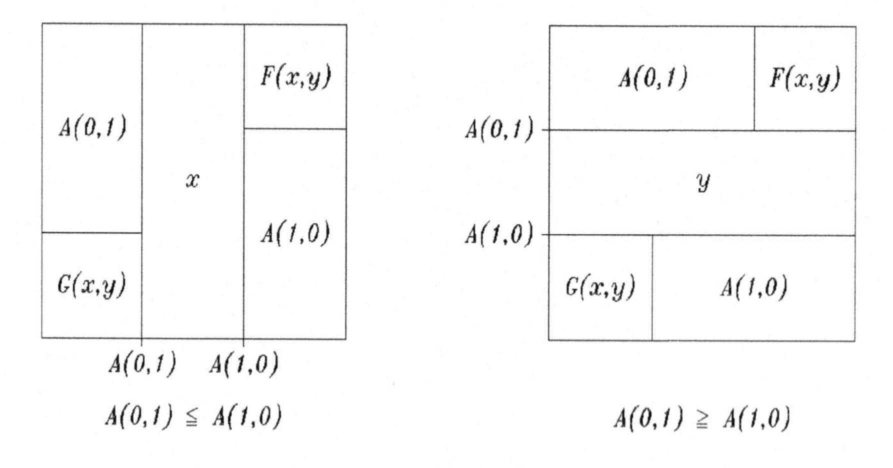

Figure 2.7.2. The functions A in Theorem 2.7.10.

if and only if $p = q$, in which case we have

Corollary 2.7.11 *Suppose that A satisfies all of the conditions given in Theorem 2.7.10. Then A is commutative if and only if for some α in I,*

$$A(x,y) = \begin{cases} G(x,y), & \text{for } (x,y) \text{ in } [0,\alpha]^2, \\ F(x,y), & \text{for } (x,y) \text{ in } [\alpha,1]^2, \\ \alpha, & \text{otherwise}, \end{cases}$$

where G and F belong to $\mathcal{G}_{C_0}[0,\alpha]$ and $\mathcal{F}_{C_0}[\alpha,1]$, respectively.

Note that $\mathrm{Min} \leq A \leq \mathrm{Max}$ if and only if $G = \mathrm{Max}$ and $F = \mathrm{Min}$.

The functions A described in Corollary 2.7.11 are frequently referred to as **nullnorms** (see Chapter 10 of [Klement, Mesiar and Pap (2000)] for further details and references to the literature).

The results of this section, as well as those of Sections 2.1 and 2.4, may be viewed as different answers to the questions posed in the second part of Hilbert's Fifth Problem.

2.8 Related functional equations

In this section we consider the relationships between the associativity equation and two other prominent functional equations – the translation equation and the bisymmetry equation.

The **translation equation**

$$\Phi(\Phi(x,u),v) = \Phi(x,u+v) \tag{2.8.1}$$

has a long and distinguished history, principally through its connection with one-dimensional dynamical systems and continuous iteration semigroups.

Consider a function $f : J \to J$, where $J = [a,b] \subseteq \mathbb{R}$, and its iterates f^n, n in $\mathbb{N}$, which here, for convenience, we shall denote by f_n. We wish to extend the iteration index from $\mathbb{N}$ to $\mathbb{R}^+$ so that

$$f_1 = f, \quad f_u \circ f_v = f_{u+v} \tag{2.8.2}$$

for all u, v in $\mathbb{R}^+$. If this is indeed possible, the resulting iteration semigroup $\{f_u\}$, u in $\mathbb{R}^+$, is called a **flow**. In this case the function $\Phi : J \times \mathbb{R}^+ \to J$ defined by

$$\Phi(x,u) = f_u(x) \tag{2.8.3}$$

is a solution of (2.8.1). Conversely, each solution Φ of (2.8.1) yields a flow via (2.8.3). (When f is invertible, the corresponding problem of extension to an iteration group follows on replacing $\mathbb{R}^+$ by $\mathbb{R}$.)

Now it is well known that, under fairly weak monotonicity and continuity assumptions, solutions of (2.8.1) admit the representation

$$\Phi(x,u) = \varphi^{-1}(\varphi(x) + u), \tag{2.8.4}$$

where $\varphi : J \to \mathbb{R}$ is continuous and strictly monotonic; and conversely, for each such φ, the function Φ defined by (2.8.4) is a solution of (2.8.1). For details, consult [Aczél (1966)].

Observe that, upon combining (2.8.3) and (2.8.4), the question of whether a given f can be embedded in a flow depends on the existence of solutions of the Abel equation $\varphi(f(x)) = \varphi(x) + 1$.

Of particular interest are the so-called **continuous flows**, those for which the function Φ in (2.8.3) is continuous in its second place, or equivalently, for which the **trajectories** $h_x : \mathbb{R}^+ \to J$, defined by $h_x(u) = f_u(x)$, are continuous for each x in J. For an extensive treatment of results on continuous flows up to 1980, consult [Targonski (1981)].

The structure of continuous one-dimensional flows in $\mathbb{R}^+$ (with some very mild additional restrictions) was completely determined by A. Sklar [1987]. A key part of the characterization involves flows derived from semigroups:

Theorem 2.8.1 *Suppose that Φ belongs to $\mathcal{F}_{Ar}(J)$ or $\mathcal{G}_{Ar}(J)$ and that φ is an inner additive generator of Φ. For each u in $\mathbb{R}^+$ define $f_u : J \to J$ via*

$$f_u(x) = \Phi(x, \varphi^{(-1)}(u)),$$

i.e., $f_u(x) = \varphi^{(-1)}(\varphi(x) + u)$. Then $\{f_u\}$ is a continuous flow on $\mathbb{R}^+$.

In other words, a continuous flow can be obtained from any such Φ by slicing it into sections and rescaling via $\varphi^{(-1)}$.

One further remark: upon replacing addition in (2.8.1) by a two-place function G, i.e.,

$$\Phi(\Phi(x,u),v) = \Phi(x, G(u,v)),$$

a simple calculation yield that G must be associative, provided that Φ is strictly monotonic in its second place.

Necessary and sufficient conditions for the embeddability of a function $f : I \to I$ into an iteration semigroup via $T(f^m(x), f^n(x)) = f^{\gamma(m,n)}$, where T is a strict t-norm, have been obtained in [Ger (1996)].

We now turn to the bisymmetry equation. Detailed treatments of this and other related functional equations (e.g., autodistributivity) may be found in [Aczél (1996); Aczél and Dhombres (1989)].

A function $B : J^2 \to J$, $J = [a, b] \subseteq \mathbb{R}$, is **bisymmetric** if

$$B[B(x,y), B(z,w)] = B[B(x,z), B(y,w)], \tag{2.8.5}$$

for all x, y, z, w in J.

In the presence of additional conditions, it is easy to relate bisymmetry and associativity.

Theorem 2.8.2 *Let $B : J^2 \to J$ be given.*

(a) If B is bisymmetric and has an identity in J, then B is associative and commutative.

(b) If B is associative and commutative, then B is bisymmetric.

Proof. (a) Let e denote the identity; associativity follows at once upon setting $z = e$ in (2.8.5), and commutativity upon setting $x = w = e$.

(b) We have

$$\begin{aligned} B[B(x,y), B(z,w)] &= B[B(B(x,y),z),w] \\ &= B[B(x, B(y,z)), w] = B[B(x, B(z,y)), w] \\ &= B[B(B(x,z),y),w] = B[B(x,z), B(y,w)]. \qquad \square \end{aligned}$$

The following examples on I show that there are no further implications among these properties. Let a = associative, b = bisymmetric, c = commutative, e = has identity, and let $'$ denote negation.

(i) a, b, c, e': $B(x,y) = 0$;
(ii) a, b, c', e': $B(x,y) = x$;
(iii) a', b, c, e': $B(x,y) = x^2y^2$;
(iv) a', b, c', e': $B(x,y) = x^2y$;
(v) a', b', c', e: Example A.1.2 (put $x = w = \frac{7}{8}, y = \frac{3}{4}, z = 1$);
(vi) a', b', c, e: Example A.1.1.

Continuous and strictly monotonic bisymmetric functions admit representation via one-place functions (see [Aczél (1966)]).

Theorem 2.8.3 *If $B : [a,b]^2 \to [a,b]$ is continuous on $[a,b]^2$, strictly monotonic in each place on $(a,b)^2$, and bisymmetric, then there exists a continuous and strictly increasing function $\beta : [a,b] \to \mathbb{R}$ and numbers p, q, r with $pq > 0$ such that for all (x,y) in $[a,b]^2$,*

$$B(x,y) = \beta^{-1}(p\beta(x) + q\beta(y) + r). \tag{2.8.6}$$

In the other direction, if for some given β, p, q, r as above, the right-hand side of (2.8.6) is well-defined, then the function B is bisymmetric.

Note that when p and q are positive (resp., negative), B is strictly increasing (resp., decreasing) in each place. Note also that B is commutative if and only if $p = q$ and that B has an identity e if and only if $p = q = 1$ and $r = -\beta(e)$.

We cite two elementary examples: (1) if $\beta = -\log$ on I, $p = 2, q = 1$, and $r = 0$, then $B(x,y) = x^2y$ on I^2; (2) if $\beta(x) = \frac{1}{x}$ on $\mathbb{R}^+$, $p = q = 1$, and $r = -1 = -\beta(1)$, then $B(x,y) = P/(\Sigma - P)$ on $\mathbb{R}^+$, with identity 1.

An especially important class of bisymmetric functions is obtained by taking p and q positive with $p + q = 1$, and $r = 0$: Specifically, a **weighted quasi-arithmetic mean** on $J = [a,b]$ is a function $M : J^2 \to J$ of the form

$$M(x,y) = f^{-1}(pf(x) + (1-p)f(y)), \tag{2.8.7}$$

where p is in (0,1) and the generator f is a continuous and strictly monotonic function from J onto some interval K. (Note that M is well-defined since $pK + (1-p)K = K$.) When $p = \frac{1}{2}$, M is called a **quasi-arithmetic mean**. This class has been extensively studied as "averaging" functions, i.e., functions that are continuous, strictly increasing in each place, and – as is easily shown – satisfy

$$\mathrm{Min}(x,y) \le M(x,y) \le \mathrm{Max}(x,y).$$

See [Hardy, Littlewood and Pólya (1952), Chapter III; Aczél (1966); Aczél and Dhombres (1989)]. We list the most prominent examples.

(1) The weighted arithmetic mean $M(x,y) = px + (1-p)y$ on $J \subseteq \mathbb{R}$; $f(x) = x$.
(2) The weighted geometric mean $M(x,y) = x^p y^{1-p}$ on $J \subseteq \mathbb{R}^+$; $f(x) = \log x$.

(3) The weighted harmonic mean $M(x,y) = \frac{xy}{(1-p)x+py}$ on $J \subseteq \mathbb{R}^+$; $f(x) = \frac{1}{x}$.

(4) The weighted root-mean powers $M_k(x,y) = (px^k + (1-p)y^k)^{1/k}$, $k > 0$, on $J \subseteq \mathbb{R}^+$; $f(x) = x^k$.

A useful observation concerning the representation (2.8.7) is that, for a fixed p, f and f_1 generate the same mean if and only if $f_1(x) = kf(x) + c$ for some constants $k \neq 0$ and c. In fact, putting $u = f(x)$, $v = f(y)$, and $\varphi = f_1 \circ f^{-1}$ into the equation $M_1 = M$ yields

$$\varphi(pu + (1-p)v) = p\varphi(u) + (1-p)\varphi(v). \tag{2.8.8}$$

The statement is then an immediate consequence of the following well-known result [Hardy, Littlewood and Pólya (1952), §3.7].

Lemma 2.8.4 *If φ is a continuous function, equation (2.8.8) is equivalent to Jensen's equation (1.2.4), whose general continuous solution is $\varphi(u) = ku + c$.*

In conclusion, we address the question: For a given weighted quasi-arithmetic mean M on I, do there exist distinct t-norms T_1 and T_2 for which $M(T_1, T_2)$ is a t-norm? The answer is affirmative when the generator f of M is infinite-valued at $x = 0$. In view of the preceding observation, we may assume that f is strictly decreasing on I, with $f(0) = \infty$ and $f(1) = 0$.

Theorem 2.8.5 *Let M be a weighted quasi-arithmetic mean on I, with weight p and generator f as above. For $\alpha > 0$, define t_α on I by*

$$t_\alpha(x) = \log(1 + \alpha f(x)),$$

and let T_α be the strict t-norm generated by t_α. Then

$$M(T_\alpha, T_\beta) = T_{p\alpha+(1-p)\beta}. \tag{2.8.9}$$

Proof. It is easy to see that t_α satisfies the conditions for an additive generator of a strict t-norm and that $t_\alpha^{-1}(u) = f^{-1}\left(\frac{e^u - 1}{\alpha}\right)$.

Hence,

$$\begin{aligned} f(T_\alpha(x,y)) &= \frac{1}{\alpha}[\exp(t_\alpha(x) + t_\alpha(y)) - 1] \\ &= \frac{1}{\alpha}[(1 + \alpha f(x))(1 + \alpha f(y)) - 1] \\ &= \alpha f(x)f(y) + f(x) + f(y), \end{aligned}$$

so that

$$\begin{aligned} pf(T_\alpha(x,y)) &+ (1-p)f(T_\beta(x,y)) \\ &= (p\alpha + (1-p)\beta)f(x)f(y) + f(x) + f(y) \\ &= f(T_{p\alpha+(1-p)\beta}(x,y)), \end{aligned}$$

which is equivalent to (2.8.9). □

In particular, the weighted harmonic mean yields Family (2.6.3) and the weighted geometric mean yields Family (2.6.9).

This result can be extended, in the obvious way, from I to $[a, b]$ and elements of $\mathcal{F}_{St}[a, b]$ when $f(a) = \infty$. A parallel result can be formulated for $\mathcal{G}_{St}[a, b]$ when $f(b)$ is infinite. For instance, the arithmetic mean on $[1, \infty]$ yields Example A.9.5 of Appendix A.

However, the question remains completely open for t-norms when $f(0)$ is finite, e.g., when M is a weighted arithmetic mean. More generally, it is not known whether a convex combination of distinct t-norms is ever a t-norm. (Some special cases are treated in [Tomás (1987)].)

Finally, we mention that representation theorems for solutions of the **generalized associativity equation**

$$F(G(x,y),z) = H(x, K(y,z))$$

have been obtained by many authors. See [Aczél (1966)] for the basics and [Maksa (2000)] for recent results.

Chapter 3

Functional equations involving t-norms

3.1 Simultaneous associativity

The primary focus of this section is on the functional equation

$$T(x, y) + S(x, y) = x + y, \qquad \text{for all } x, y \text{ in } I, \tag{3.1.1}$$

for t-norms and s-norms. A solution is a pair (T, S) satisfying (3.1.1) with T in $\mathcal{T}$ and S in $\mathcal{S}$. This equation arises naturally in the study of the associativity of certain binary operations on probability distributions [Frank (1975), (1991)] and in a number of other settings as well. Here we shall present the complete solution of (3.1.1) and some related equations, investigate extensions, and discuss some consequences of these results.

The first thing to note about (3.1.1) is that, since S is non-decreasing, T satisfies the Lipschitz condition (1.4.8). Hence T must be a copula, say C, and S must be $C^{\wedge}$, the associated dual copula of C which is given by (1.4.9), i.e.,

$$C^{\wedge}(x, y) = x + y - C(x, y). \tag{3.1.2}$$

Thus, in view of Lemma 1.4.6, our aim is to determine all solution pairs $(C, C^{\wedge})$ for which C is an associative copula and the dual copula $C^{\wedge}$ is also (simultaneously) associative.

To begin, consider the one-parameter family (2.6.5) of strict copulas, viz.,

$$C_\alpha(x, y) = -\frac{1}{\alpha}\log\left[1 + \frac{(e^{-\alpha x} - 1)(e^{-\alpha y} - 1)}{e^{-\alpha} - 1}\right], \quad \alpha \neq 0 \tag{3.1.3a}$$

and its limits

$$C_{-\infty} = W, \qquad C_0 = P, \qquad C_\infty = \text{Min}. \tag{3.1.3b}$$

A straightforward calculation yields that

$$x + y - C_\alpha(x, y) = 1 - C_\alpha(1 - x, 1 - y),$$

i.e., that

$$C_\alpha^{\wedge} = C_\alpha^*; \tag{3.1.4}$$

and since the associativity of C_α implies that C_α^* is associative, it follows that, for each α in $\mathbb{R}$, the pair (C_α, C_α^*) is a solution of (3.1.1).

The remarkable and surprising fact is that these pairs (C_α, C_α^*) and pairs $(C, C^\wedge)$ for which C is an ordinal sum of C_α's are the only solutions. For we have:

Theorem 3.1.1 *For T in $\mathcal{T}$ and S in $\mathcal{S}$, the pair (T, S) is a solution of (3.1.1) if and only if*

(a) $T = C_\alpha$, where C_α is given by (3.1.3) for some α in $\mathbb{R}$, and $S = C_\alpha^*$;

or

(b) T is an ordinal sum of C_α's and $S = T^\wedge$.

Note that in case (b), $S = T^*$ if and only if the idempotent set of T is symmetric with respect to the point $x = 1/2$. Note also that without the monotonicity asssumptions on T and S, other associative solutions of (3.1.1) exist; see Example A.5.1 of Appendix A.

Theorem 3.1.1 is the main subject of [Frank (1979)]. Its proof is intricate and too long to be reproduced in its entirety here. Instead, we shall present a synopsis of the key steps in the argument, referring the reader to [Frank (1979)] for the details.

The first thing to note is that if the pair $(C, C^\wedge)$ satisfies (3.1.1), then C and $C^\wedge$ have identical idempotent sets. From this and Theorem 2.4.3 it follows at once that an "ordinal sum" $(C, C^\wedge)$ of pairs $(C_\alpha, C_\alpha^\wedge)$ satisfies (3.1.1) if and only if each pair of corresponding summands does. This observation reduces the search to Archimedean solutions.

The next preliminary step is to show that the pair (W, W^*) is the only non-strict Archimedean solution. To this end, suppose that the non-strict pair $(C, C^\wedge)$ is a solution and that C is generated by t, with $t(0) = 1$. Let $x_0 = \sup\{x : C(x, x) = 0\}$.
Then $x_0 = t^{(-1)}(1/2)$, and $0 < x_0 \le 1/2$ since t is convex. Next, for all x in $[0, x_0]$

$$x + 2x_0 - C(x, 2x_0) = C^\wedge(x, 2x_0) = C^\wedge(x, C^\wedge(x_0, x_0))$$

$$= C\hat{}(C\hat{}(x, x_0), x_0) = C\hat{}(x + x_0, x_0)$$
$$= x + 2x_0 - C(x + x_0, x_0),$$

whence, since t and $t^{(-1)}$ are strictly decreasing on I,

$$C(x, 2x_0) = C(x + x_0, x_0) > 0.$$

Hence, for all x in $[0, x_0]$,

$$t(x) - t(x + x_0) = t(x_0) - t(2x_0).$$

Consequently, the slope of the segment joining the points $(x, t(x))$ and $(x_0 + x, t(x_0 + x))$ is equal to the slope of the segment joining the points $(x_0, t(x_0))$ and $(2x_0, 2t(x_0))$. It follows that t, being convex, must be linear on $[0, 2x_0]$, and further that $2x_0 = 1$, since $t(0) = 1$ and $t(x_0) = 1/2$ together imply that $t(2x_0) = 0 = t(1)$. Thus t is linear on I and $C = W$ (see Theorem 10 of [Frank (1975)]).

It thus remains to find all strict solutions, i.e., to solve the functional equation

$$t^{-1}(t(x) + t(y)) + s^{-1}(s(x) + s(y)) = x + y, \tag{3.1.5}$$

for pairs of convex additive generators with $t(0) = s(1) = \infty$. Here the key is to reduce (3.1.5) to several second-order differential equations.

Delicate analysis yields that t and s must be differentiable on (0,1), that t' and s' must be strictly increasing (hence continuous) on (0,1), and that t and s are related as follows:

(a) If $t'(1-) = 0$, then $t's = -k_1$.
(b) If $s'(0+) = 0$, then $ts' = k_2$.
(c) If $t'(1-) < 0$, then $s(x) = -k_3 \log[1 - t'(1-)/t'(x)]$.
(d) If $s'(0+) > 0$, then $t(x) = -k_4 \log[1 - s'(0+)/s'(x)]$.

Here k_1, k_2, k_3, k_4 are positive constants. It now follows that both t' and s' must be continuously differentiable on (0,1). Consequently, in each of the four possible pairs of cases – viz. (a) and (b), (a) and (d), (c) and (b), (c) and (d) – it is legitimate to differentiate one of the equations and substitute into the other in order to arrive at a second-order differential equation involving either t or s. Specifically, we obtain:

(1) If $t'(1-) = 0 = s'(0+)$, then

$$tt'' = k(t')^2, \quad k > 0.$$

(2) If $t'(1-) = 0 < s'(0+)$, then

$$ss'' = ks'[s' - s'(0+)], \quad k > 0.$$

(3) If $t'(1-) < 0 = s'(0+)$, then

$$tt'' = kt'[t' - t'(1-)], \quad k > 0.$$

(4) If $t'(1-) < 0 < s'(0+)$, then

$$(e^t - 1)t'' = t'[t'(1-) - t']e^t.$$

(In case (4), a long argument is needed to show that $k_3 t'(1-) = -k_4 s'(0+)$.)

In the first three cases it can be shown that all generator solutions of the differential equation in question are such that either $t(0) < \infty$ or $s(1) < \infty$, and thus are not solutions of (3.1.5). For case (1), simple integrations and case splitting suffice; for cases (2) and (3), an additional argument showing that $k = 1$ is needed. Finally for the equation in case (4), integration yields that

$$t^{-1}(u) = \int \frac{du}{t'(1-) - b(e^u - 1)}, \quad b > 0.$$

If $b = -t'(1-)$, then $t(x) = -\log x$. Otherwise, letting $y = e^u - 1$, integrating, inverting, imposing the condition $t(0) = \infty$, and setting $t'(1-) = \alpha/(1 - e^{-\alpha})$, yields the solutions

$$t_\alpha(x) = -\log \frac{e^{-\alpha x} - 1}{e^{-\alpha} - 1}, \quad \alpha \neq 0, \tag{3.1.6}$$

i.e., the generators of the family (3.1.3a).

Note: Employing the representation theorem for Archimedean functions to reduce a functional equation in multiplace functions to a more tractable equation in one-place functions is a recurrent theme in this chapter.

The family of copulas (3.1.3), which was in fact discovered in the course of solving the equation (3.1.1), has a number of remarkable properties. As a consequence, it has come to play an important (and totally unforeseen) role in a variety of settings. Foremost among these are statistical modeling and the study of questions of statistical dependence.

In [Kimeldorf and Sampson (1975a)] the authors stipulate that in order to be useful in certain statistical applications a one-parameter family $\{C_\theta\}$,

θ in $[a, b] \subseteq \mathbb{R}$, of copulas (which they at the time referred to as *uniform representations*) should, at the very least, have the following properties:

(i) The family includes the Fréchet-Hoeffding bounds, $W = C_a$ and $\mathrm{Min} = C_b$, and the copula P of independence – either explicitly or as limiting cases.

(ii) For each fixed (x, y), the family $\{C_\theta(x, y)\}$ is continuous in θ.

(iii) For each θ in the open interval (a, b), C_θ is absolutely continuous, i.e., has a density.

In addition, they and others frequently also require that:

(iv) The family $\{C_\theta\}$ is stochastically ordered, viz., positively directed, so that $C_{\theta_1} \leq C_{\theta_2}$ whenever $\theta_1 \leq \theta_2$.

(v) For each θ in (a, b), the support of C_θ is I^2.

A number of families having the properties (i)-(v) are known, e.g., the copulas of bivariate normal distributions and the copulas of the Plackett family. However, among all these, to date, the family (3.1.3) is the only known family of elementary and associative copulas. Herein, in its many symmetry properties, and in the fact that many properties of associative copulas can be expressed as properties of their generators, lies its attractiveness and utility (see, e.g., [Nelsen (1986); Genest (1987); Hutchinson and Lai (1990); Joe (1997)]). A systematic discussion and many further references may be found in the book [Nelsen (1999)]. Here we only present some of the salient results that relate directly to simultaneous associativity.

For any copula C, let C^L, C^R, and $\overline{C}$ be the copulas defined by

$$C^L(x, y) = y - C(1 - x, y), \tag{3.1.7a}$$

$$C^R(x, y) = x - C(x, 1 - y), \tag{3.1.7b}$$

$$\overline{C}(x, y) = x + y - 1 + C(1 - x, 1 - y). \tag{3.1.8}$$

Elementary calculations yield

$$(C^L)^R = (C^R)^L = \overline{C}, \tag{3.1.9}$$

$$(C^L)^L = (C^R)^R = C, \tag{3.1.10}$$

$$(C^\wedge)^* = \overline{C}. \tag{3.1.11}$$

Theorem 1.4.4(e) yields the following stochastic interpretation of these copulas: If X and Y are continuous random variables with $C_{XY} = C$ and if f and g are decreasing functions, then

$$C_{f(X),Y} = C^L, \quad C_{X,g(Y)} = C^R, \quad C_{f(X),g(Y)} = \overline{C}.$$

The copula $\overline{C}$ is also the **survival copula** associated with C since, on combining (1.4.10) and (3.1.11), we obtain:

$$\overline{H}(u,v) = \overline{C}(\overline{F}_X(u), \overline{F}_Y(v)), \tag{3.1.12}$$

where $\overline{F}_X$ and $\overline{F}_Y$ are the survival functions and $\overline{H}$ is the joint survival function, i.e.,

$$\overline{F}_X(u) = Pr(X > u), \quad \overline{F}_Y(v) = Pr(Y > v),$$

and

$$\overline{H}(u,v) = Pr(X > u, Y > v).$$

We also note that for continuous, symmetric random variables X and Y, the identity $C_{XY} = \overline{C}_{XY}$ characterizes the property of **radial symmetry** (see [Nelsen (1999)]).

It readily follows from (3.1.11) that $\overline{C}$ is associative if and only if $C^{\wedge}$ is associative. Consequently Theorem 3.1.1 yields

Theorem 3.1.2 *The copula C and its survival copula $\overline{C}$ are simultaneously associative if and only if C is a member C_α of the family (3.1.3) or an ordinal sum of members of this family.*

Note that for any member C_α of the family (3.1.3),

$$\overline{C}_\alpha = (C_\alpha^{\wedge})^* = (C_\alpha^{\wedge})^{\wedge} = C_\alpha,$$

and that $\overline{C} = C$ for ordinal sums of members of this family if and only if the idempotent set of C is symmetric about $x = 1/2$. Moreover,

$$C_\alpha^R = C_\alpha^L = C_{-\alpha}. \tag{3.1.13}$$

Commutativity and the fact that $\overline{C}_\alpha = C_\alpha$ together imply that, for each α in $\mathbb{R}$, the density γ_α of C_α is symmetric about the lines $y = x$ and $y = 1 - x$; and (3.1.13) implies that

$$\gamma_\alpha(1-x, y) = \gamma_\alpha(x, 1-y) = \gamma_{-\alpha}(x,y)$$

(see [Genest (1987); Nelsen (1999)]).

For the next result we require two lemmas, the first of which is easily established.

Lemma 3.1.3 *For any copula C:*

(a) If C and C^L (or C and C^R) are commutative, then $C^L = C^R$.
(b) $C^L = C^R$ if and only if $C = \overline{C}$.

Lemma 3.1.4 *If the copula C is a proper ordinal sum, say of $([a_n, b_n], C_n)$, then neither C^L nor C^R can be associative.*

Proof. Note first that the point $(1-x, y)$ belongs to the zero-set $\mathcal{Z}(C^L)$ if and only if $y = C(x, y)$, i.e., if and only if $C(x, y) = \mathrm{Min}(x, y)$ and $y \leq x$. Thus $\mathcal{Z}(C^L)$ is the reflection in the line $x = 1/2$ of the set

$$\{(x, y) \text{ in } I^2 : x \leq y\} \setminus \bigcup_{n=1}^{\infty} (a_n, b_n)^2.$$

Consequently, for each interval $[a_n, b_n]$, both the horizontal segment $[1-a_n, 1-b_n] \times \{a_n\}$ and the vertical segment $\{(1-b_n)\} \times [a_n, b_n]$ belong to the boundary curve of $Z(C^L)$. Thus this boundary curve is not the graph of a continuous decreasing function, whence (see (2.1.12)) C^L is not associative. Similarly, C^R is not associative. □

Theorem 3.1.5 *The copulas C and C^L (or C and C^R) are simultaneously associative if and only if C is a member of the family (3.1.3).*

Proof. Suppose that C and C^L are both associative. Then C and C^L are both commutative and by Lemma 3.1.3, $C = \overline{C}$. By Lemma 3.1.4, C cannot be an ordinal sum and thus, by Theorem 3.1.2, C is a member of the family (3.1.3). The converse follows directly from (3.1.13). Essentially the same argument applies when C^L is replaced by C^R. □

Corollary 3.1.6 *Let C be an associative copula. Then $C = C^L$ (or $C = C^R$) if and only if $C = P$.*

Proof. This follows immediately from Theorem 3.1.5, (3.1.13) and (3.1.3b). □

We conclude this discussion with several observations which amplify the preceding results:

(1) While (3.1.3) is the only known family of elementary, associative copulas having all of the Kimeldorf-Sampson properties, Family (2.6.1) comes close: the $\{T_\alpha\}$ satisfy (i)-(iv) for α in $[-1, \infty]$ and also (v) for α in $[0, \infty]$. Another family of copulas satisfying (i)-(v) is exhibited in Example A.5.2 of Appendix A. A method of constructing such families is given in [Genest and MacKay (1986b), p. 157].

(2) Some non-associative copulas for which $C = C^L$ are listed in Example A.5.3 of Appendix A.

(3) The statistical properties of the family (3.1.3) have been exploited by many authors. In addition to the previously cited references, we mention [Genest and Rivest (1989), (1993); Capéraà and Genest (1990); Joe (1993); Meester and MacKay (1994); Nelsen (1997)].

The functional equation (3.1.1) plays a prominent role in multivalued logics and the theory of fuzzy sets (Section 3.2) and in preference modeling. See especially [Klement (1981a), (1982a,b); Alsina (1988); Fodor and Roubens (1994b); Nguyen and Walker (2000)] and the references cited therein. Also, the corresponding functional equation

$$\tau(F, G) + \tau^{\wedge}(F, G) = F + G \tag{3.1.14}$$

for some families of triangle functions is solved in [Alsina (1986b)].

Next we consider the extension of equation (3.1.1) from I to an arbitrary interval J, i.e., solutions (F, G) of

$$F(u, v) + G(u, v) = u + v, \qquad \text{for all } u, v \text{ in } J, \tag{3.1.15}$$

with F in $\mathcal{F}(J)$ and G in $\mathcal{G}(J)$. We restrict our attention to Archimedean solutions since ordinal sums behave exactly as on I. Note that (3.1.15) is consistent with the boundary conditions (2.3.3) and (2.3.4).

For a compact interval $J = [a, b]$, the proof of Theorem 3.1.1 can be extended almost verbatim on replacing I by J. Since (2.3.13) transforms (3.1.1) to (3.1.15), the one-parameter family $\{F_\alpha\}$ of strict solutions can be obtained via (2.3.13), together with the limits (2.3.16), (2.3.17) and Min; and the corresponding family $\{G_\alpha\}$ is obtained via (2.3.18). For unbounded intervals, the proof can be adapted to yield the solutions of (3.1.15), but the situation is a bit more complicated. A similar argument reduces the Archimedean to the strict case, i.e., to the analogue of equation (3.1.5), and then to the differential equations given in (1)-(4), with the obvious changes in notation and endpoints. But now exactly one of the equations (1)-(3) yields a one-parameter family of solutions: (1) when $J = \mathbb{R}$;

(2) when $J = \mathbb{R}^+$; and (3) when $J = \mathbb{R}^-$. We omit the details and simply state the results which, although never published, have been known for some time [Frank (1981)].

(1) When $J = \mathbb{R}$, then necessarily $\alpha > 0$, and the solutions (F_α, G_α) together with their additive generators are given by

$$F_\alpha(u,v) = -\frac{1}{\alpha}\log(e^{-\alpha u} + e^{-\alpha v}), \quad f_\alpha(u) = e^{-\alpha u},$$
$$G_\alpha(u,v) = -\frac{1}{\alpha}\log(e^{\alpha u} + e^{\alpha v}), \quad g_\alpha(u) = e^{\alpha u}. \tag{3.1.16}$$

Note that the F_α are the natural extensions from $\mathbb{R}^+$ to $\mathbb{R}$ of the Wiener-Shannon laws (2.6.27).

(2) When $J = \mathbb{R}^+$, then again $\alpha > 0$, and the solutions (F_α, G_α) are the composition laws (2.6.28) and (2.6.30).

(3) When $J = \mathbb{R}^-$, the solutions (F'_α, G'_α), $\alpha > 0$, are obtained from the composition laws (2.6.28) and (2.6.30) via

$$F'_\alpha(u,v) = -G_\alpha(-u,-v) \text{ and } G'(u,v) = -F_\alpha(-u,-v). \tag{3.1.17}$$

On intervals of the form $[c,\infty]$ or $[-\infty,c]$, where $-\infty < c < \infty$, the solutions are obtained from the solutions in (2) or (3) above via a simple translation. For instance, on $[c,\infty]$, we have $F''_\alpha(u,v) = c + F_\alpha(u-c, v-c)$, where F_α is given by (2.6.28).

Note that in all cases, $F_\infty = \text{Min}$ and $G_\infty = \text{Max}$ and that at least one of F_0, G_0 is discontinuous on the boundary of J^2. For example, in case (2), $G_0(u,v) = u + v$, so that $F_0(u,v) = 0$ for all (u,v) in $[0,\infty)^2$.

As a final generalization of (3.1.1), we consider the extension of (3.1.15) to the equation

$$\Phi(F(u,v), G(u,v)) = \Phi(u,v), \qquad \text{for all } (u,v) \text{ in } [a,b]^2, \tag{3.1.18}$$

where Φ is a given two-place function defined on $[a,b]^2$. For any such Φ, we seek pairs of functions (F,G) satisfying (3.1.18) with F in $\mathcal{F}[a,b]$ and G in $\mathcal{G}[a,b]$ (see [Frank (1981)]).

Observe that when Φ is commutative, (Min, Max) is a solution. (There may be no others, for example when $\Phi(u,v) = (u^2+v^2)/uv$. In this case, putting $u = v$ in (3.1.18) and setting $q(u) = F(u,u)/G(u,u)$, we find that $[q(u)]^2 + 1 = 2q(u)$, so that $q \equiv 1$ and $F(u,u) = G(u,u) = u$ for all u.)

We shall assume that Φ is expressible in the form

$$\Phi(u, v) = \psi(\phi(u) + \phi(v)), \tag{3.1.19}$$

where ϕ is continuous and strictly increasing from $[a, b]$ onto an interval J and ψ is invertible on the interval $J + J$. (Thus, when $\phi = j_{[a,b]}$ and $\psi = j_{[2a,2b]}$, (3.1.18) is (3.1.15) on $[a, b]$.) Note that Φ belongs to $\mathcal{F}_{st}[a, b]$ (resp., $\mathcal{G}_{st}[a, b]$) when $J = \mathbb{R}^-$ (resp., $\mathbb{R}^+$) and $\psi = \phi^{-1}$. We confine our attention to Archimedean solutions (F, G), since the situations for ordinal sums are somewhat complicated and of only minor interest.

Under these conditions on Φ, equation (3.1.18) is easily reduced to (3.1.15): from (3.1.19) we have that (3.1.18) is equivalent to

$$\phi(F(u, v)) + \phi(G(u, v)) = \phi(u) + \phi(v),$$

which, with $w = \phi(u)$ and $z = \phi(v)$, is in turn equivalent to

$$F^{\square}(w, z) + G^{\square}(w, z) = w + z, \qquad \text{for all } w, z \text{ in } J, \tag{3.1.20}$$

where

$$F(u, v) = \phi^{-1}[F^{\square}(\phi(u), \phi(v))],$$

$$G(u, v) = \phi^{-1}[G^{\square}(\phi(u), \phi(v))], \tag{3.1.21}$$

and $F^{\square}$, $G^{\square}$ belong to $\mathcal{F}_{Ar}(J)$ and $\mathcal{G}_{Ar}(J)$, respectively. Observe that by means of a suitable translation and dilation, J can always be adjusted to be one of the basic intervals $I, \mathbb{R}, \mathbb{R}^+$, or $\mathbb{R}^-$.

To solve (3.1.18) it therefore suffices to take the appropriate family $(F_\alpha^{\square}, G_\alpha^{\square})$ of solutions of (3.1.20) and transform them to (F_α, G_α) via (3.1.21). These families are given in (3.1.3a), (3.1.16), (2.6.28), (2.6.30) and (3.1.17). Note that if the $(F_\alpha^{\square}, G_\alpha^{\square})$ are generated by $(f_\alpha^{\square}, g_\alpha^{\square})$, then the solutions (F_α, G_α) of (3.1.18) are generated by

$$f_\alpha = f_\alpha^{\square} \circ \phi \qquad \text{and} \qquad g_\alpha = g_\alpha^{\square} \circ \phi. \tag{3.1.22}$$

We present several illustrative examples:

(a) $$T(x, y)S(x, y) = xy, \qquad \text{for all } x, y \text{ in } I.$$

Here $\Phi(x, y) = xy$, so that $\phi = \log$, $J = \mathbb{R}^-$ and $f_\alpha^{\square}(w) = 1 - e^{-\alpha w}$. Therefore, $t_\alpha(x) = 1 - x^{-\alpha}$, and (T, S) is an Archimedean solution if and only if $T = T_\alpha$ and $S = S_\alpha$ for some $\alpha > 0$, where

$$T_\alpha(x, y) = (x^{-\alpha} + y^{-\alpha} - 1)^{-1/\alpha}$$

and

$$S_\alpha(x, y) = (x^\alpha + y^\alpha - x^\alpha y^\alpha)^{-1/\alpha}.$$

Note that the T_α are from Family (2.6.1) and the S_α^* are from Family (2.6.6).

(b) $$F(u, v)G(u, v) = uv, \qquad \text{for all } u, v \text{ in } \mathbb{R}^+.$$

Again $\phi = \log$, but now $J = \mathbb{R}$, so that $f_\alpha^\square(w) = e^{-\alpha w}$ and $f_\alpha(u) = -u^\alpha$. The solutions (F_α, G_α), $\alpha > 0$, are thus

$$F_\alpha(u, v) = (u^{-\alpha} + v^{-\alpha})^{-1/\alpha}$$

and

$$G_\alpha(u, v) = (u^\alpha + v^\alpha)^{1/\alpha}.$$

These are the hyperbolic laws (2.6.26) and the Minkowski laws (2.6.29), respectively.

(c) $$\frac{T(x, y)S(x, y)}{T(x, y) + S(x, y)} = \frac{xy}{x + y}, \qquad \text{for all } (x, y) \text{ in } I^2\backslash\{(0, 0)\}.$$

Here we may take $\phi(x) = 1 - x^{-1}$ and $\psi(w) = (2 - w)^{-1}$ so that $J = \mathbb{R}^-$. Then $t_\alpha(x) = 1 - \exp(\alpha/x - \alpha)$, and we obtain the solutions (T_α, S_α), $\alpha > 0$, where T_α belongs to Family (2.6.22).

(d) For $\Phi(x, y) = xy - (x + y)/xy$ for (x, y) in $(0, 1]^2$ we take $\phi(x) = -\log(x^{-1} - 1)$, so that $J = \mathbb{R}$, and we obtain the generators $t_\alpha(x) = (x^{-1} - 1)^\alpha$, $\alpha > 0$, yielding Family (2.6.12).

Note: The solution of (3.1.18) was used in [Sánchez-Soler (1988)] to solve the corresponding generalization of (3.1.14) for triangle functions.

We conclude this section with a brief discussion of the problem that was the genesis of the functional equation (3.1.1). For each copula C, let $\sigma_{C,+}$ be the binary operation on the space Δ of probability distributions defined for all F, G in Δ via the Lebesgue-Stieltjes integral

$$\sigma_{C,+}(F, G)(x) = \iint\limits_{u+v \le x} d(F(u), G(v)), \quad -\infty < x < \infty. \tag{3.1.23}$$

With the notation of (1.4.5), for any pair of real random variables X and Y, we have that

$$\sigma_{C_{XY},+}(F_X, F_Y) = F_{X+Y}.$$

The family $\{\sigma_{C,+}\}$ thus extends convolution ($C = P$) to arbitrary pairs of (dependent) random variables.

The following question arises naturally in the study of the triangle inequality for probabilistic metric spaces: For which copulas C is the operation $\sigma_{C,+}$ a triangle function? This question, in turn, leads to the more general question: For which C is $\sigma_{C,+}$ associative? The complete answers to both of these questions were given in [Frank (1975)]; and the arguments in that paper were greatly simplified in [Frank (1979)] via the use of Theorem 3.1.1.

It is not difficult to show that if $\sigma_{C,+}$ is associative, then C must be associative and a solution of the distributivity equation

$$C(x + y - C(x, y), z) = C(x, z) + C(y, z) - C(C(x, y), z). \tag{3.1.24}$$

Moreover, if C is a solution of (3.1.24), then $C^{\hat{}}$ must be associative. Hence, if $\sigma_{C,+}$ is associative, then $(C, C^{\hat{}})$ must be a solution of (3.1.1). For Archimedean C, setting $x = y = z = 1/2$ and calculating shows that neither W nor any C_α given by (3.1.3a) is a solution of (3.1.24); nor can these copulas be components of any ordinal sum. Since P and Min do satisfy (3.1.3a), these observations yield:

Theorem 3.1.7 *An associative copula is a solution of (3.1.24) if and only if $C = P$, $C =$ Min, or C is an ordinal sum, each summand of which is P.*

In [Frank (1975)] it is further proved that $\sigma_{C,+}$ is indeed associative whenever C is as given in Theorem 3.1.7. These results are extended from addition to a wide class of operations L on random variables via corresponding operations $\sigma_{C,L}$ in [Frank (1991)]. Further details and background information concerning the operations $\sigma_{C,L}$ may be found in [Schweizer and Sklar (1983), (2005)] and [Schweizer (1991)].

3.2 n-duality

The function $1 - j_I$ is an involution on I, and it is because of this property that the function B^* in (1.3.12) is an s-norm whenever B is a t-norm. It is also clear that the same is true for a larger class of involutions.

Definition 3.2.1 A **strict involution** on I is a function n with domain I which is continuous, strictly decreasing, and such that $n(0) = 1$ and $n \circ n = j_I$ or, equivalently, $n = n^{-1}$.

Clearly $n(1) = 0$ and Ran $n = I$. Note also that if T is in $\mathcal{T}_{Ar} \backslash \mathcal{T}_{St}$, then the boundary curve of $\mathcal{Z}(T)$ is the graph of a strict involution (see 2.1.12).

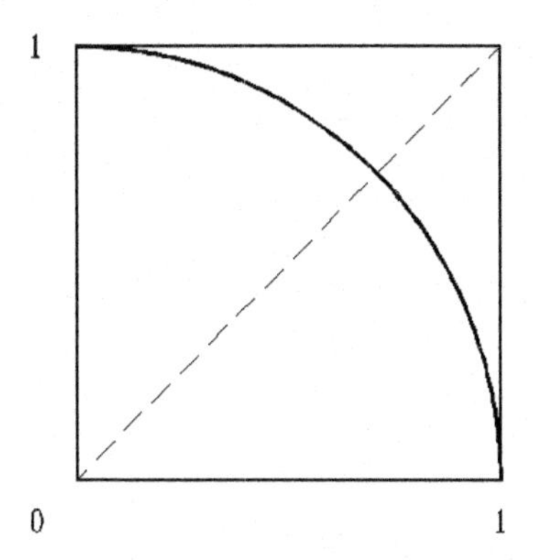

Figure 3.2.1. The graph of a strict involution.

In various works on multivalued logic and the related parts of lattice theory, strict involutions are sometimes referred to as **strong negations** [Trillas (1979), (1980)].

If T is a t-norm and n is a strict involution on I, then by way of Theorem 2.3.3, the function S defined on I^2 by

$$S(x, y) = n(T(n(x), n(y))) \tag{3.2.1}$$

is an s-norm which is in $\mathcal{S}_{Co}$, $\mathcal{S}_{St}$ or $\mathcal{S}_{Ar}$, whenever T is in $\mathcal{T}_{Co}$, $\mathcal{T}_{St}$ or $\mathcal{T}_{Ar}$, respectively. Note also that since $n = n^{-1}$, (3.2.1) yields

$$T(x, y) = n(S(n(x), n(y))). \tag{3.2.2}$$

If a t-norm T and an s-norm S are related as in (3.2.1) we shall say that they are **n-dual**. Note that any t-norm T and its associated s-norm T^* are $(1 - j_I)$-dual.

Theorem 3.2.2 *Let T be in $\mathcal{T}_{Ar}$ and S in $\mathcal{S}_{Ar}$; let t and s be inner additive generators of T and S, respectively; and let n be a strict involution.*

Then T and S are n-dual if and only if there exists an $\alpha > 0$ such that

$$s \circ n = \alpha t. \tag{3.2.3}$$

Proof. From (3.2.1) and the representation theorem, for all x, y in I we have

$$t^{(-1)}(t(x) + t(y)) = (n \circ s^{(-1)})(s(n(x)) + s(n(y))).$$

But from (2.1.1) and (2.1.16) it follows that $n \circ s^{(-1)} = (s \circ n)^{(-1)}$. Thus $s \circ n$ is also an inner additive generator of T, and (3.2.3) follows from Corollary 2.2.6. □

Next, from (3.2.3) and the fact that $n = n^{-1}$, we have:

Corollary 3.2.3 *There exists a strict involution n such that T and S are n-dual if and only if there exists an $\alpha > 0$ such that*

$$s^{(-1)} \circ (\alpha t) = t^{(-1)} \circ \left(\frac{1}{\alpha} s\right). \tag{3.2.4}$$

If T and S are both non-strict, so that $t(0) < \infty$ and $s(1) < \infty$, and if (3.2.3) holds, then necessarily $\alpha = s(1)/t(0)$. Thus, if, in this case, there exists a strict involution n such that T and S are n-dual, then n is unique and is given by

$$n = s^{(-1)} \circ \left(\frac{s(1)}{t(0)} t\right). \tag{3.2.5}$$

Such an involution n may not exist. For example, if W and the s-norm $S(x, y) = \text{Min}\left(\sqrt{x^2 + y^2}, 1\right)$ were n-dual, then $\alpha = 1$ and (3.2.5) yields $n(x) = \sqrt{1 - x}$, which is not a strict involution.

If one of T, S is strict and the other non-strict, then clearly there is no α such that (3.2.3) holds, and hence no n for which T and S are n-dual.

The situation is much more interesting when T and S are both strict. The following results are due to P. Garcia and L. Valverde [1989].

Theorem 3.2.4 *Let T be in $\mathcal{T}_{St}$, with additive generator t, and let S be in $\mathcal{S}_{St}$, with additive generator s. Suppose T and S are n-dual. Let $\alpha > 0$ be any constant whose existence is guaranteed by Corollary 3.2.3, and let*

$$D(T, S) = \left\{\beta : s^{-1} \circ \frac{\alpha}{\beta} t = t^{-1} \circ \frac{\beta}{\alpha} s\right\}. \tag{3.2.6}$$

Then $D(T,S)$ is a relatively closed subgroup of the multiplicative group of positive reals.

Proof. By Corollary 3.2.3, the identity 1 is in $D(T,S)$. Next suppose that β and γ are in $D(T,S)$ so that $\beta \neq 0$, $\gamma \neq 0$, and that

$$s^{-1} \circ \frac{\alpha}{\beta} t = t^{-1} \circ \frac{\beta}{\alpha} s \text{ and } s^{-1} \circ \frac{\alpha}{\gamma} t = t^{-1} \circ \frac{\gamma}{\alpha} s.$$

Then

$$s^{-1} \circ \frac{\alpha}{\beta} t \circ t^{-1} \circ \frac{\gamma}{\alpha} s = t^{-1} \circ \frac{\beta}{\alpha} s \circ s^{-1} \circ \frac{\alpha}{\gamma} t,$$

whence

$$s^{-1} \circ \frac{\gamma}{\beta} s = t^{-1} \circ \frac{\beta}{\gamma} t,$$

which, by Corollary 3.2.3, yields

$$s^{-1} \circ \frac{\gamma}{\beta} s \circ s^{-1} \circ \alpha t = t^{-1} \circ \frac{\beta}{\gamma} t \circ t^{-1} \circ \frac{1}{\alpha} s,$$

i.e.,

$$s^{-1} \circ \frac{\gamma}{\beta} \alpha t = t^{-1} \circ \frac{\beta}{\gamma} \cdot \frac{1}{\alpha} s.$$

Thus β/γ is in $D(T,S)$, and so $D(T,S)$ is a multiplicative subgroup of $\mathbb{R}^+ \backslash \{0\}$. Lastly, relative closure of $D(T,S)$ in $\mathbb{R}^+ \backslash \{0\}$ is an immediate consequence of the continuity of the generators and their inverses. □

The group $D(T,S)$ in (3.2.6) is independent both of the particular choice of α (if there is a choice) and of the choice of the generators t and s. First, to show independence of the choice of α, suppose there is a $\gamma \neq \alpha$ such that

$$s^{-1} \circ \gamma t = t^{-1} \circ \frac{1}{\gamma} s,$$

and let $D'(T,S) = \left\{ \beta : \, s^{-1} \circ \frac{\gamma}{\beta} \, t = t^{-1} \circ \frac{\beta}{\gamma} \, s \right\}$.

If β is in $D'(T,S)$, then we have

$$s^{-1} \circ \frac{1}{\beta} s = s^{-1} \circ \frac{\gamma}{\beta} t \circ t^{-1} \circ \frac{1}{\gamma} s = t^{-1} \circ \frac{\beta}{\gamma} s \circ s^{-1} \circ \gamma t = t^{-1} \circ \beta t,$$

whence, using the fact that $s^{-1} \circ \alpha t = t^{-1} \circ \frac{1}{\alpha} t$, we have

$$s^{-1} \circ \frac{\alpha}{\beta} t = t^{-1} \circ \frac{\beta}{\alpha} s.$$

Thus β is in $D(T,S)$ and $D'(T,S) \subseteq D(T,S)$. Similarly one obtains that $D(T,S) \subseteq D'(T,S)$.

Next, suppose that τ and σ are also generators of T and S, respectively. Then there exist $a, b > 0$ such that $\tau = at$ and $\sigma = bs$. Thus for any $\alpha > 0$,

$$s^{-1} \circ \alpha t = \sigma^{-1} \circ \frac{b}{a} \alpha \tau$$

and

$$t^{-1} \circ \frac{1}{\alpha} s = \tau^{-1} \circ \frac{a}{b} \cdot \frac{1}{\alpha} \sigma.$$

It follows that

$$D_{\tau,\sigma}(T,S) = \left\{ \beta : \sigma^{-1} \circ \frac{b}{a} \cdot \frac{\alpha}{\beta} \tau = \tau^{-1} \circ \frac{a}{b} \cdot \frac{\beta}{\alpha} \sigma \right\} = D(T,S).$$

Note that if we set

$$A_{t,s}(T,S) = \left\{ \alpha : \ s^{-1} \circ \alpha t = t^{-1} \circ \frac{1}{\alpha} s \right\}, \tag{3.2.7}$$

then

$$A_{\tau,\sigma}(T,S) = \frac{b}{a} A_{t,s}(T,S).$$

Observe that the sets $D(T,S)$ and $A_{t,s}(T,S)$ have the same cardinality.

We now take a closer look at the group $D(T,S)$. Since it is a relatively closed subgroup of the multiplicative group $\mathbb{R}^+ \setminus \{0\}$, it must, if non-empty, be one of the following:

(i) $D(T,S) = \{1\}$;
(ii) $D(T,S) = (0, \infty)$;
(iii) $D(T,S) = \{\rho^k : \ k \text{ in } \mathbb{Z}\}$ for some ρ in $(0,1)$.

Thus, for any given T and S, there are either no, one, countably many, or uncountably many strict involutions n such that T and S are n-dual. The set of these involutions is given by

$$\left\{ n_\alpha : n_\alpha = s^{-1} \circ \frac{1}{\alpha} t, \text{ for some } \alpha \text{ in } A_{t,s}(T,S) \right\}.$$

All possibilities occur, as the following examples show.

Example 3.2.5 $D(T,S) = \phi$.

For the strict t-norm T generated by $t(x) = 2(x^{-2} - 1)$ and the strict s-norm S generated by $s(x) = (1-x)^{-1} - 1$, we have

$$\left(s^{-1} \circ (\alpha t)\right)(x) = \frac{2\alpha(1-x^2)}{x^2 + 2\alpha(1-x^2)},$$

and

$$\left(t^{-1} \circ \left(\frac{1}{\alpha} s\right)\right)(x) = \left(1 + \frac{x}{2\alpha(1-x)}\right)^{1/2};$$

a simple computation shows that for all $\alpha > 0$ equality fails for $x = \frac{1}{2}$. Thus T and S cannot be n-dual for any n.

Example 3.2.6 $D(T,S) = \{1\}$.

For the strict t-norm T generated by $t(x) = \log(2x^{-1} - 1)$ and the strict s-norm S generated by $s(x) = \log[(1+x)/(1-x)]$, we have

$$\left(s^{-1} \circ (\alpha t)\right)(x) = \frac{(2-x)^\alpha - x^\alpha}{(2-x)^\alpha + x^\alpha},$$

and

$$\left(t^{-1} \circ \left(\frac{1}{\alpha} s\right)\right)(x) = \frac{2(1-x)^\alpha}{(1-x)^\alpha + (1+x)^\alpha}.$$

Thus (3.2.4) holds if and only if

$$\frac{(2-x)^\alpha - x^\alpha}{(2-x)^\alpha + x^\alpha} = \frac{2(1-x)^\alpha}{(1-x)^\alpha + (1+x)^\alpha}. \tag{3.2.8}$$

Substituting $x = \frac{1}{2}$ into (3.2.8) yields that $\alpha = 1$ is the unique solution, and consequently that $n(x) = 1 - x$ is the unique strict involution for which T and S are n-dual.

Example 3.2.7 $D(T,S) = (0,\infty)$.

For the strict t-norm T generated by $t(x) = x^{-1} - 1$, the strict s-norm S generated by $s(x) = (1-x)^{-1} - 1$, and for all $\alpha > 0$, we have

$$\left(s^{-1} \circ (\alpha t)\right)(x) = \left(t^{-1} \circ \left(\frac{1}{\alpha} s\right)\right)(x) = \frac{\alpha(1-x)}{x + \alpha(1-x)} = n_\alpha(x).$$

Thus T and S are n_α-duals for all α in $(0,\infty)$.

Example 3.2.8 $D(T,S) = \{2^k : k \text{ in } \mathbb{Z}\}$.

Let $T = P$, with generator $t(x) = -\log x$, and let S be the s-norm generated by $s(x) = f(-\log x)$, where $f : \mathbb{R}^+ \to \mathbb{R}^+$ is defined on $\left(0, \frac{1}{2}\right]$ by

$$f(x) = 2^m\left(\frac{3}{2} - 2^m x\right), \text{ for } x \text{ in } \left[2^{-m-1}, 2^{-m}\right], \; m = 1, 2, \ldots,$$

and by $f = f^{-1}$ elsewhere. Then f is well-defined and (3.2.4) holds if and only if

$$\frac{1}{\alpha} f(-\log x) = f(\alpha(-\log x)). \tag{3.2.9}$$

In [Garcia and Valverde (1982)] it was shown that (3.2.9) is satisfied if and only if $\alpha = 2^k$ for some k in $\mathbb{Z}$.

The negations in Example 3.2.7 are the only strong negations that are rational functions. This follows from

Theorem 3.2.9 *A strict involution n is a rational function if and only if, for some $\lambda < 1$ and for all x in I,*

$$n(x) = \frac{1-x}{1-\lambda x}.$$

Proof. Suppose that $n(x)$ is rational and let $n(z)$ be the analytic continuation of $n(x)$ from I to the extended complex plane $\overline{\mathbb{C}}$. Then $n(n(z)) = z$ for all z in $\overline{\mathbb{C}}$, whence $n(z)$ is univalent. But it is well known that any univalent analytic function that maps $\overline{\mathbb{C}}$ onto $\overline{\mathbb{C}}$ must be bilinear [Levinson and Redheffer (1970), Section 5.7]. Thus $n(z) = (az+b)/(cz+d)$ for all z in $\overline{\mathbb{C}}$ and the rest follows from the boundary conditions, $n(0) = 1$ and $n(1) = 0$, and the fact that $n(x)$ is real and strictly decreasing on I. The converse is trivial. □

An interesting family of non-differentiable strict involutions has been studied in [Mayor and Torrens (1992), (1994); Mayor (1994)]. These are the solutions n_λ, $0 < \lambda < 1$, of the functional equation

$$n(\lambda x) = \lambda + (1-\lambda)n(x), \quad \text{for all } x \text{ in } I,$$

and are given by $n_\lambda(x) = f_\lambda(1-f_\lambda^{-1}(x))$, where f_λ is the continuous solution of De Rham's system of functional equations

$$\begin{cases} f\left(\frac{x}{2}\right) = \lambda f(x), \\ f\left(\frac{1+x}{2}\right) = \lambda + (1-\lambda) f(x), \end{cases}$$

for all x in I (see [De Rham (1956)]).

In classical set theory, each subset A of a given set X can be identified with its associated characteristic function χ_A, where

$$\chi_A(x) = \begin{cases} 0, \; x \text{ not in } A, \\ 1, \; x \text{ in } A. \end{cases}$$

Indeed, given $\{\chi_A : A \subseteq X\}$ one may define $A \cap B$, $A \cup B$ and A^c via

$$\begin{aligned} \chi_{A\cap B} &= \operatorname{Min}(\chi_A, \chi_B), \\ \chi_{A\cup B} &= \operatorname{Max}(\chi_A, \chi_B), \\ \chi_{A^c} &= (1-j) \circ \chi_A. \end{aligned}$$

In the theory of fuzzy sets and in fuzzy logic, following L.A. Zadeh [1965], a **fuzzy subset** of a given set X is a mapping A from X to I. A Boolean-like structure may be placed on the set of all $A : X \to I$ by defining

$$\begin{aligned} (A \vee B)(x) &= \operatorname{Min}(A(x), B(x)), \\ (A \wedge B)(x) &= \operatorname{Max}(A(x), B(x)), \\ A^c(x) &= ((1-j) \circ A)(x) = 1 - A(x); \end{aligned}$$

and this may be further generalized by setting

$$\begin{aligned} (A \vee B)(x) &= T(A(x), B(x)), \\ (A \wedge B)(x) &= S(A(x), B(x)), \\ A^c(x) &= (n \circ A)(x) = n(A(x)), \end{aligned}$$

where (T, S, n) is a **De Morgan triple**, i.e., n is a strict involution, T is a t-norm, S is an s-norm, and T, S are n-duals. Note that the duality $n(T(x,y)) = S(n(x), n(y))$ corresponds to the classical De Morgan law $(A \cap B)^c = A^c \cup B^c$. In the literature such De Morgan triples are also referred to as **generalized logical connectives** (see, e.g., [Klement, Mesiar and Pap (2000); Trillas, Alsina and Terricabras (1995)]).

In this context, given certain desirable properties of $\wedge$ and $\vee$, three typical situations arise: (1) the usual pair Min-Max is the unique solution; (2) several De Morgan triples satisfy the given conditions; (3) there exist solutions which cannot be n-duals. We illustrate each case with one example.

Theorem 3.2.10 *Let T be a continuous t-norm and S a continuous s-norm. If S and T are n-dual for all strict involutions n on I, then $T = Min$ and $S = Max$.*

Proof. We will show that $S(x_0, x_0) = x_0$ for all x_0 in $(0,1)$. To this end, let n_1 and n_2 be two strict involutions on I such that x_0 is a fixed point, i.e., $n_1(x_0) = n_2(x_0) = x_0$ but $n_1(x) < n_2(x)$ whenever x is in $I \backslash \{0, 1, x_0\}$. Then $n_1(S(x_0, x_0)) = T(n_1(x_0), n_1(x_0)) = T(n_2(x_0), n_2(x_0)) = n_2(S(x_0, x_0))$, whence $S(x_0, x_0)$ belongs to $\{0, 1, x_0\}$. Since $S \geq$ Max, $S(x_0, x_0)$ is either x_0 or 1. If $S(x_0, x_0) = 1$, then $T(x_0, x_0) = 0$ which implies that $T = Z$ and contradicts the continuity of T. Thus $S(x_0, x_0) = x_0$, whence $S =$ Max and $T =$ Min. □

The second situation is illustrated by the following theorem, which is also of considerable interest in its own right.

Theorem 3.2.11 *Let (T, S, n) be a De Morgan triple, where T and S are continuous. Suppose that the* **orthomodularity** *property holds, i.e., that*

$$T(x, S(y, n(x))) = y, \tag{3.2.10}$$

for all x, y in I with $y \leq x$. Then there exists a continuous, strictly decreasing function t from I to $\mathbb{R}^+$ such that $t(1) = 0$, $t(0) < \infty$, and

$$T(x, y) = t^{(-1)}(t(x) + t(y)), \tag{3.2.11}$$

$$n(x) = t^{-1}(t(0) - t(x)), \tag{3.2.12}$$

$$S(x, y) = t^{-1}[Max(t(x) + t(y) - t(0), 0)]. \tag{3.2.13}$$

Proof. Substituting $y = 0$ into (3.2.10), we have that $T(x, n(x)) = 0$ for all x in I. Since $n(x) > 0$ for all x in $(0,1)$, it follows from (2.4.5) that T cannot be a proper ordinal sum and therefore has no interior idempotents (see Figure 2.4.2). Thus T is a non-strict Archimedean t-norm and admits a representation of the form (3.2.11). Now let x be fixed and suppose that $z > n(x)$. Define $f : I \to I$ by $f(u) = S(u, n(x))$. Since f is continuous,

$f(1) = 1$, and

$$f(0) = n(x) < z \leq 1 = n(0) = n(T(x, n(x))) \\ = S(n(x), x) = S(x, n(x)) = f(x),$$

there exists $u_0 > 0$ such that $z = f(u_0)$. Then, from (3.2.10) we obtain

$$T(x, z) = T(x, f(u_0)) = T(x, S(u_0, n(x))) = u_0 > 0.$$

Thus n is the boundary curve of $\mathcal{Z}(T)$ and (3.2.12) holds. Finally, since S is the n-dual of T, (3.2.13) follows. □

A De Morgan triple (T, S, n) which admits the representations (3.2.11), (3.2.12) and (3.2.13) is called a **Lukasiewicz triple**.

Turning to the third situation, recall that the identity

$$(A \cap B) \cup (A \cap B^c) = A \tag{3.2.14}$$

holds in every Boolean algebra. For the fuzzy sets analogue, it is therefore natural to look for triples (T, S, n) satisfying the functional equation

$$S(T(x, y), T(x, n(y))) = x.$$

This equation was solved in [Alsina (1985b)] and its consequences in fuzzy logic were explored in [Alsina (1996)]. We first establish the following lemma which is of intrinsic interest.

Lemma 3.2.12 *If T is an associative binary operation on I which is non-decreasing in its second place and satisfies the equation*

$$T(x, y) + T(x, 1 - y) = x \tag{3.2.15}$$

for all x, y in I, then $T = P$.

Proof. Substituting $x = 0$ into (3.2.15) yields

$$T(0, y) + T(0, 1 - y) = 0,$$

whence, since $T \geq 0$, $T(0, y) = 0$ for all y in I. Next, the substitution $y = \frac{1}{2}$ in (3.2.15) yields $T(x, \frac{1}{2}) = \frac{x}{2}$ for all x in I. Thus

$$T(x, 0) = T\left(x, T\left(0, \frac{1}{2}\right)\right) = T\left(T(x, 0), \frac{1}{2}\right) = \frac{1}{2}T(x, 0),$$

i.e., $T(x, 0) = 0$ for all x in I. Now an easy induction shows that

$$T\left(x, \sum_{i=1}^{n} y_i/2^i\right) = \sum_{i=1}^{n} xy_i/2^i \tag{3.2.16}$$

holds whenever $y_1, y_2, \ldots, y_n$ are in $\{0, 1\}$. Hence, for any x in I and any dyadic rational q in I,

$$T(x, q) = xq.$$

Since T is non-decreasing in its second place and the dyadic rationals are dense in I, taking left- and right-hand limits yields $T(x, y) = xy$ for all x, y in I. □

Theorem 3.2.13 *If T is a continuous t-norm, S a continuous s-norm, and n a strict involution on I, then the general solution of the functional equation*

$$S(T(x, y), T(x, n(y))) = x \tag{3.2.17}$$

is given by

$$S(x, y) = s^{(-1)}(s(x) + s(y)), \tag{3.2.18}$$

$$n(x) = s^{-1}(1 - s(x)), \tag{3.2.19}$$

$$T(x, y) = s^{-1}(s(x) \cdot s(y)), \tag{3.2.20}$$

where s is a continuous and strictly increasing function from I to I, with $s(0) = 0$ and $s(1) = 1$.

Proof. Letting $x = 1$ in (3.2.17) yields $S(y, n(y)) = 1$ for all y in I. Thus, as in the proof of Theorem 3.2.11, it follows that S is an non-strict Archimedean s-norm. The representation theorem now yields (3.2.18), where, without loss of generality, s is the additive generator of S for which $s(1) = 1$. Consequently, since x is in I and since $s \circ s^{(-1)}(u) = u$ whenever $0 \le u \le s(1) = 1$, it follows that (3.2.17) is equivalent to

$$s(T(x, y)) + s(T(x, n(y))) = s(x). \tag{3.2.21}$$

Let x_0 denote the unique fixed point of the strict involution n. Then setting $y = x_0$ in (3.2.21), we have

$$2s(T(x, x_0)) = s(x);$$

and substituting $x = x_0$ in (3.2.21), we obtain

$$s(T(x_0, x)) + s(T(x_0, n(x))) = s(x_0).$$

Hence we have

$$\frac{1}{2}s(x) + \frac{1}{2}s(n(x)) = s(x_0),$$

which on setting $x = 0$ yields $s(x_0) = \frac{1}{2}s(1) = \frac{1}{2}$, so that

$$s(x) + s(n(x)) = 1,$$

and consequently

$$n(x) = s^{-1}(1 - s(x)),$$

i.e., n has the form (3.2.19).

Now let B be the binary operation on I given by

$$B(x, y) = s\left(T(s^{-1}(x), s^{-1}(y))\right).$$

It is obvious that B is a t-norm. Next (3.2.19) and (3.2.21) yield

$$\begin{aligned} B(x, y) + B(x, 1 - y) &= s\big(T(s^{-1}(x), s^{-1}(y))\big) + s\big(T(s^{-1}(x), s^{-1}(1 - y))\big) \\ &= s\big(T(s^{-1}(x), s^{-1}(y))\big) + s\big(T(s^{-1}(x), n(s^{-1}(y)))\big) \\ &= s(s^{-1}(x)) = x. \end{aligned}$$

Thus by Lemma 3.2.12, $B = P$, whence T has the form (3.2.20), and the theorem is proved. □

Note that the t-norm T in (3.2.20) and the s-norm S in (3.2.18) cannot be n-dual for any strict involution n because T is strict and S is not. The triples (T, S, n) given in Theorem 3.2.13 are often called **normal triples**.

A generalization of (3.2.17) was solved independently in [Fodor and Roubens (1994a)] and [Alsina (1997)]. Let T_1 and T_2 be continuous t-norms, S a continuous s-norm, and n a strict involution on I. Then, in part as a consequence of Theorem 3.1.1, the general solution of the equation

$$S(T_1(x, y), T_2(x, n(y))) = x$$

is as follows: S and n are given by (3.2.18) and (3.2.19); T_1 and T_2 are given by

$$T_1(x, y) = s^{-1}(T_\alpha(s(x), s(y))),\ T_2(x, y) = s^{-1}(T_{-\alpha}(s(x), s(y))),$$

where T_α is any member of Family (2.6.5), $-\infty \leq \alpha \leq \infty$, i.e., the family (3.1.3). Note that $\alpha = 0$ yields (3.2.20).

The proof mimics that of Theorem 3.2.13, except that (3.2.15) is replaced by

$$T_1(x, y) + T_2(x, 1 - y) = x,$$

and, since $T_2 = T_1^R$, Lemma 3.2.12 is replaced by Theorem 3.1.5 together with (3.1.13).

The following result appears in [Alsina, Trillas and Valverde (1983)].

Theorem 3.2.14 *Given a t-norm T, an s-norm S, and a strong negation n, the triple (T, S, n) satisfies* **Kleene's inequality**

$$T(x, n(x)) \leq S(y, n(y)), \text{ for all } x, y \text{ in } I.$$

Proof. Let z be the fixed point of n. Then for all x, y in I,

$$T(x, n(x)) \leq \text{Min}(x, n(x)) \leq z \leq \text{Max}(y, n(y)) \leq S(y, n(y)). \qquad \square$$

The preceding discussion shows that the study of multivalued logics and the theory of fuzzy sets leads naturally to functional equations involving associative functions. In the next portion of this section, we present a number of additional results along these lines but omit the proofs.

A pair (T, n), where T is a t-norm and n a strong negation, satisfies **the principle of non-contradiction** (PNC) if $T(x, n(x)) = 0$, for all x in I; and a pair (S, n), where S is an s-norm and n a strong negation, satisfies **the law of the excluded middle** (LEM) if $S(x, n(x)) = 1$, for all x in I. Obviously (see Theorem 3.2.11), if the pair (T, n) satisfies the PNC, then T is representable in the form $T(x, y) = t^{(-1)}(t(x)+t(y))$ and $n(x) = t^{-1}(t(0)-t(x))$, where $t(0) < \infty$; and the pair (S, n) satisfies the LEM if and only if $S(x, y) = s^{(-1)}(s(x) + s(y))$ and $n(x) = s^{-1}(s(1) - s(x))$, where $s(1) < \infty$.

The fundamental notion of implication has been studied as a relation in [Zadeh (1978), (1979)] and as a connective in [Bandler and Kohout (1980); Gaines (1976); Trillas and Valverde (1981); Alsina and Trillas (1984)]. In this second approach, for given fuzzy subsets A and B of X, the implication $A \to B$ is the fuzzy subset of X defined by $(A \to B)(x) = \mathcal{I}(A(x), B(x))$, where $\mathcal{I}$ is specified by

Definition 3.2.15 An **implication function** is a two-place function $\mathcal{I} : I^2 \to I$ which satisfies the following properties:

(i) **Negation**: $\mathcal{I}(x, 0) = n(x)$, where n is a strict involution on I;
(ii) **Continuity**: $\mathcal{I}$ is continuous on I^2;
(iii) **Monotonicity**: $\mathcal{I}(x', y') \leq \mathcal{I}(x, y)$ whenever $x \leq x'$, and $y \leq y'$;
(iv) **Falsity Principle**: $\mathcal{I}(0, x) = 1$;
(v) **Neutrality Principle**: $\mathcal{I}(1, x) = x$;
(vi) **Principle of Contrapositivity**: $\mathcal{I}(x, y) = \mathcal{I}(n(y), n(x))$;
(vii) **Exchange Principle**: $\mathcal{I}(x, \mathcal{I}(y, z)) = \mathcal{I}(y, \mathcal{I}(x, z))$.

Implication functions are easily characterized by way of the following theorem [Aczél (1966); Alsina and Trillas (1984)].

Theorem 3.2.16 *A mapping $\mathcal{I}$ from I^2 to I which is continuous in its first place, and which satisfies both the Exchange Principle and the boundary conditions $\mathcal{I}(0,0) = 1$, $\mathcal{I}(1,0) = 0$, has the form $\mathcal{I}(x, y) = S(f(x), y)$, where $f : I \to I$ is continuous with $f(0) = 1$, $f(1) = 0$, and S is a binary operation on I which is commutative, is associative, and has identity 0. Consequently, every implication function has the form*

$$\mathcal{I}(x, y) = S(n(x), y),$$

where S is a continuous s-norm and n is a strict involution.

From the representation theorem for continuous Archimedean s-norms, implication functions have the explicit form

$$\mathcal{I}(x, y) = s^{(-1)}\left(s(n(x)) + s(y)\right).$$

In the particular case $S = W^*$, this yields the classical Luckasiewicz implication

$$\mathcal{I}(x, y) = \mathrm{Min}(1 - x + y, 1).$$

In this context we also have the identity

$$S(n(x), y) = n(T(n(n(x)), n(y))) = n(T(x, n(y))),$$

which is the well known **linguistic implication** function corresponding to the operator $p \to q = \neg(p \& \neg q)$.

If, for a given an implication function $\mathcal{I}(x, y) = S(n(x), y)$, where S is Archimedean, we define

$$x \leq_{\mathcal{I}} y \text{ if and only if } \mathcal{I}(x, y) \geq \mathcal{I}(x, x),$$

then $\leq_{\mathcal{I}}$ is the usual order relation on I. In general $\leq_{\mathcal{I}}$ is a quasi-order. Moreover the **"Modus Ponens"**,

$$T(x, \mathcal{I}(x, y)) \leq_{\mathcal{I}} y, \tag{3.2.22}$$

is satisfied by the t-norm T which is the n-dual of $S(x, y) = \mathcal{I}(n(x), y)$. The inequality (3.2.22) has been used in Intuitionistic Logic as well as in Quantum Logic [Hardegree (1981)].

We now consider other possible approaches to implication functions in multivalued logic.

In extending the formalism of Intuitionistic Logic, several implication operators have been introduced [Gaines (1976); Pavelka (1979); Bandler and Kohout (1980); Trillas and Valverde (1981)]. In this context we recall the following:

Definition 3.2.17 A **residuated implication** (briefly, an R-implication) is a binary operation R_T on I given by

$$R_T(x, y) = \sup\{z \text{ in } I : T(z, x) \leq y\},$$

where T is a continuous t-norm.

An R-implication satisfies the first five properties of Definition 3.2.15 and (I, T, R_T) is always a residuated lattice. The case $R_{\text{Min}}(x, y) = 1$ if $x \leq y$ and $R_{\text{Min}}(x, y) = y$ otherwise, was studied in [Bandler and Kohout (1980)]; the case $R_P(x, y) = \text{Min}(1, x/y)$ was introduced in [Gaines (1976)]; and the Lukasiewicz implication is just $R_W(x, y) = \text{Min}(1 - x + y, 1)$. In general, for an Archimedean t-norm T generated by t, we have

$$R_T(x, y) = t^{(-1)}(t(y) - t(x)).$$

Following the approach to implication given in Quantum Logic [Bodiou (1964); Hardegree (1981)], Trillas and Valverde [1985] introduced the notion of QM-implication in multivalued logics.

Definition 3.2.18 A function Q from I^2 to I is a **QM-implication** if there exists a De Morgan triple (T, S, n) such that

$$Q(x, y) = S(n(x), T(x, y)). \tag{3.2.23}$$

Among the many properties of QM-implications are the following:

Theorem 3.2.19 *If Q is the QM-implication determined by the De Morgan triple (T, S, n), then $n(x) = Q(x, 0)$ if and only if $Q(x, y) =$*

$Max(n(x), y)$. In this case, the QM-implication is just an R-implication which satisfies Modus Ponens.

Definition 3.2.20 For a given set X and t-norm T, a **T-indistinguishability operator** is a function E from X^2 to I which satisfies the following conditions for all x, y, z in X:

(i) **Reflexivity**: $E(x,x) \geq \lambda$, for some λ in I;
(ii) **Symmetry**: $E(x,y) = E(y,x)$;
(iii) **T-transitivity**: $T(E(x,y), E(y,z)) \leq E(x,z)$.

Such operators appear in many different contexts. Thus the **probabilistic relations** introduced by Menger [1951a] are P-indistinguishability operators with $\lambda = 1$; the **similarity relations** of Zadeh [1975, 1979] are Min-operators; the **likeness relations** of Ruspini [1982] are W-operators. A general study of these operators has been carried out in [Trillas (1982)] and [Trillas and Valverde (1984)]. For example, **"equivalence"** ($\leftrightarrow$) among fuzzy sets A, B in I^X may be defined as the fuzzy set $(A \leftrightarrow B)$ in I^X which is given by

$$(A \leftrightarrow B)(x) = E(A(x), B(x)),$$

where E is a T-indistinguishability operator. We cite one result [Trillas (1982); Trillas and Valverde (1984); Valverde (1984)]:

Theorem 3.2.21 *Let E be mapping from X^2 to I and let T be a continuous t-norm. Then E is a T-indistinguishability operator with $\lambda = 1$ if and only if there exists a family $\{h_\alpha\}_{\alpha in A}$ of fuzzy subsets of X such that*

$$E(x,y) = inf_{\alpha in A} R_T(Max(h_\alpha(x), h_\alpha(y)) \mid Min(h_\alpha(x), h_\alpha(y))),$$

where R_T is the R-implication associated with T.

T-indistinguishability operators also arise naturally in the theory of generalized metric spaces. Thus when T is strict, the T-transitivity of the relation E is equivalent to the triangle inequality for $d(x,y) = t(E(x,y))$ [Alsina and Trillas (1985)].

An extensive exposition of the preceding topics is the subject of the monograph [Klement, Mesiar and Pap (2000)]; see especially Chapters 10-12 and the references cited therein. Also, several topics – in particular, De Morgan triples and implications – are treated in some detail in Chapters 5 and 6 of the textbook [Nguyen and Walker (2000)].

To conclude this section, we consider a family of distances on I induced by copulas. These metrics are motivated by generalizations of the classical symmetric difference of sets.

Given a t-norm T, define $d_T : I^2 \to I$ by

$$d_T(x,y) = \begin{cases} T^*(x,y) - T(x,y), & x \neq y, \\ 0, & x = y. \end{cases} \tag{3.2.24}$$

It is immediate that d_T is symmetric and

$$d_{\text{Min}}(x,y) = |x-y| \leq T^*(x,y) - T(x,y) = d_T(x,y).$$

The following argument [Alsina (1984b)] establishes the triangle inequality.

Theorem 3.2.22 *If the t-norm T is a copula, then d_T is a metric.*

Proof. For distinct x, y, z in I we have

$$\begin{aligned} T(1-x,1-z) &\leq \text{Min}(1-x,1-z), \\ T(x,z) &\leq \text{Min}(x,z), \\ T(1-z,1-y) - T(1-x,1-y) &\leq \text{Max}(x-z,0), \\ T(z,y) - T(x,y) &\leq \text{Max}(z-x,0). \end{aligned}$$

Adding these four inequalities, we obtain $T(1-x,1-z) + T(x,z) + T(1-z,1-y) - T(1-x,1-y) + T(z,y) - T(x,y) \leq 1$, which in turn implies

$$d_T(x,y) \leq d_T(x,z) + d_T(z,y). \qquad \square$$

An interesting open question is the following: if T is a continuous t-norm such that d_T is a metric, is T necessarily a copula?

There are binary operations T on I which are not associative copulas but for which d_T can be a well-defined metric. For example, consider

$$T_1(x,y) = \begin{cases} x\left(2 - \frac{1}{y}\right), & \frac{1}{2} \leq y \leq 1 \text{ and } x \leq y \leq 1, \\ y\left(2 - \frac{1}{x}\right), & \frac{1}{2} \leq x \leq 1 \text{ and } y \leq x \leq 1, \\ 0, & \text{otherwise.} \end{cases}$$

Then T_1 is a continuous, non-Lipschitz binary operation on I, and d_{T_1} is a metric. One may even obtain metrics d_T from discontinuous operations, e.g., by letting

$$T_2(x,y) = \begin{cases} 0, & \text{if } x+y \le 1, \\ xy, & \text{if } x+y > 1. \end{cases}$$

Another way to build metrics d_T is due to W.F. Darsow. Let ϕ be a continuous, non-decreasing function from I to I such that $\phi(x) = 0$ for x in $\left[0, \frac{1}{2}\right]$ and $\phi(1) = 1$, and define

$$T_3(x,y) = \mathrm{Min}(x,y) \cdot \phi(\mathrm{Max}(x,y)).$$

Then d_{T_3} is a metric if and only if $2 - \frac{1}{x} \le \phi(x)$ for $\frac{1}{2} \le x \le 1$. Note, however, that in each of the examples T_1, T_2 and T_3, the operations are non-associative.

3.3 Simple characterizations of Min

We begin by noting once again that Min is the unique t-norm such that $T(x,x) = x$ for all x in I, i.e., such that $\delta_T = j_I$. A more interesting property of Min is given by

Theorem 3.3.1 *Let T be non-decreasing in each place on I^2 and such that $T(x,1) = T(1,x) = x$ for all x in I, i.e., suppose that T satisfies (1.3.1b) and (1.3.2). Then*

$$\sup_{u+v=2x} T(u,v) = x \tag{3.3.1}$$

for all x in I if and only if $T = $ Min.

Proof. Clearly $\sup_{u+v=2x} \mathrm{Min}(u,v) = x$. In the other direction, fix z in I and choose ϵ so that $0 < 2\epsilon < z$. By hypothesis,

$$\sup_{u+v=2(z-\epsilon)} T(u,v) = z - \epsilon.$$

Thus there exist u_0, v_0 such that $u_0 + v_0 = 2z - 2\epsilon$ and

$$(z-\epsilon) - T(u_0, v_0) < \epsilon.$$

Suppose $u_0 > z$. Then $v_0 < z - 2\epsilon < u_0$, whence

$$z - 2\epsilon < T(u_0,v_0) \le \mathrm{Min}(u_0,v_0) = v_0 < z - 2\epsilon,$$

which is a contradiction. Thus $u_0 \leq z$; and similarly, $v_0 \leq z$. Consequently,

$$z - 2\epsilon < T(u_0, v_0) \leq T(z, z) \leq z,$$

which implies that $T(z, z) = z$, and $T = \text{Min}$. □

Theorem 3.3.1 was first established in [Alsina (1981)] and is the crucial step in the proof of the fact that the τ_T-product (see [Schweizer and Sklar (1983), (2005), Chapter 7]) of the uniform probability distributions on $[a, b]$ and $[c, d]$ is the uniform distribution on $[a + b, b + d]$ if and only if $T = \text{Min}$.

Theorem 3.3.2 *Suppose that T satisfies (1.3.1b) and (1.3.2), and let T^* be given by (1.3.12), i.e., $T^*(x, y) = 1 - T(1 - x, 1 - y)$. Then*

$$T(x, y)T^*(x, y) = xy \tag{3.3.2}$$

for all x, y in I if and only if $T = \text{Min}$.

Proof. Clearly (3.3.2) is equivalent to

$$T(x, y)[1 - T(1 - x, 1 - y)] = xy. \tag{3.3.3}$$

First letting $x = y = u$ in (3.3.3) and then letting $x = y = 1 - u$ yields

$$T(u, u)[1 - T(1 - u, 1 - u)] = u^2$$

and

$$T(1 - u, 1 - u)(1 - T(u, u)) = (1 - u)^2.$$

Eliminating $T(1 - u, 1 - u)$ and simplifying, we have

$$[T(u, u)]^2 - 2uT(u, u) + u^2 = 0,$$

whence $T(u, u) = u$ for all u in I, and $T = \text{Min}$. The converse is trivial. □

Corollary 3.3.3 *Suppose that T satisfies (1.3.1b) and (1.3.2). Let n be a strict involution on I and let $S(x, y) = n(T(n(x), n(y)))$ be the n-dual of T. Then*

$$T(x, y) + S(x, y) - T(x, y)S(x, y) = 1 - n(x)n(y) \tag{3.3.4}$$

if and only if $n = 1 - j$, $T = \text{Min}$, and $S = \text{Max}$.

Proof. Substituting $y = 0$ in (3.3.4) yields $n(x) = 1 - x$. A simple calculation then yields that $S^*(x, y)T^*(x, y) = xy$ for all x, y in I. □

Theorem 3.3.4 *Suppose that T satisfies (1.3.1b) and (1.3.2). Then*

$$T(x, y) + S(x, y) = x + y$$

and

$$T(x, y) \cdot S(x, y) = xy$$

if and only if $T = Min$ and $S = Max$.

Proof. With $x = y$, $T(x, x)[2x - T(x, x)] = x^2$, i.e., $[T(x, x) - x]^2 = 0$, and the result immediately follows from Lemma 1.3.6. □

Finally, we state a useful result which is easily proved ([Alsina and Trillas (1987)]).

Theorem 3.3.5 *The t-norm Min is the unique t-norm that satisfies any one of the following conditions: For all x, y, z such that $x, y, z, x + z$, and $y + z$ are in I,*

(i) $T(x + z, y + z) = T(x, y) + z$;
(ii) $T(x + z, y + z) \geq T(x, y) + z$;
(iii) $T(x + z, y + z) \leq T(x, y) + z$;
(iv) $T(x + z, y + z) = T(x, y) + T(z, z)$.

Moreover, the result remains valid when Min is replaced by Max and T is replaced by an s-norm S.

3.4 Homogeneity

In this section we characterize homogeneous associative functions on intervals. To do so, we first consider the case of t-norms.

Theorem 3.4.1 *If T is a t-norm that satisfies the functional equation of homogeneity of degree k on I, for some $k > 0$, i.e., if*

$$T(\lambda x, \lambda y) = \lambda^k T(x, y), \tag{3.4.1}$$

for all x, y, and λ in I, then either $k = 1$ and $T = Min$ or $k = 2$ and $T = P$.

Proof. From (3.4.1) we obtain

$$T(T(x,x),x^k) = T(x^kT(1,1),x^k) = T(x^k,x^k) = x^{k^2}T(1,1) = x^{k^2},$$

and

$$\begin{aligned} T(x,T(x,x^k)) &= T(x,x^kT(1,x^{k-1})) = T(x,x^{2k-1}) \\ &= x^kT(1,x^{2k-2}) = x^{3k-2}. \end{aligned}$$

Since T is associative, $k^2 = 3k - 2$, i.e., $k = 1$ or $k = 2$. If $k = 1$, then $T(\lambda,\lambda) = \lambda$ for all λ so that $T = \text{Min}$. If $k = 2$, then for all x, y in (0,1] we have

$$\begin{aligned} T(x,y) &= T(\max(x,y),\min(x,y)) \\ &= (\max(x,y))^2T\left(1,\frac{\min(x,y)}{\max(x,y)}\right) = xy, \end{aligned}$$

whence $T = P$. □

Corollary 3.4.2 *Let T be a t-norm and let f be a function from I to I such that*

$$T(\lambda x,\lambda y) = f(\lambda)T(x,y), \tag{3.4.2}$$

for all x, y, and λ in I. Then either $f = j$ and $T = Min$ or $f = j^2$ and $T = P$.

Proof. From (3.4.2),

$$T(x,x) = T(x\cdot 1,x\cdot 1) = f(x)T(1,1) = f(x),$$

whence f is monotonic and satisfies the multiplicative Cauchy equation (1.2.3), since

$$f(xy) = T(xy,xy) = f(x)T(y,y) = f(x)f(y).$$

Thus f must be of the form $f(x) = x^k$ for some $k > 0$. Applying the preceding theorem yields the desired result. □

B.R. Ebanks [1998] has extended Corollary 3.4.2 to **quasi-homogeneous** t-norms, i.e., to T for which

$$\varphi(T(\lambda x,\lambda y)) = f(\lambda)\varphi(T(x,y)), \quad \text{for all } x,y,\lambda \text{ in } I,$$

where $\varphi : I \to \mathbb{R}$ is continuous and strictly monotonic and $f : I \to I$ is arbitrary.

Theorem 3.4.3 *A t-norm is quasi-homogeneous if and only if it is a member T_α of Family (2.6.1) with α in $\mathbb{R}^+$, i.e., if and only if*

$$T_\alpha(x, y) = (x^{-\alpha} + y^{-\alpha} - 1)^{-1/\alpha}, \quad \text{for } 0 < \alpha < \infty,$$

and

$$T_0 = P, \quad T_\infty = Min.$$

Here, for all α in $\mathbb{R}^+$, $f_\alpha(\lambda) = \lambda^c$ with arbitrary $c > 0$, and the φ_α are given by

$$\varphi_\alpha(x) = k(1 + x^{-\alpha})^{-c/\alpha}, \quad 0 < \alpha < \infty,$$

and

$$\varphi_0(x) = kx^{c/2}, \quad \varphi_\infty(x) = kx^c,$$

with $k \neq 0$. The proof rests in part on the general solution of the functional equation

$$t(\lambda x) = a(\lambda)t(x) + b(\lambda)$$

presented in [Aczél and Dhombres (1989), Chapter 15, Theorem 1].

For s-norms satisfying the functional equation of homogeneity, the situation is different.

Theorem 3.4.4 *Let S be an homogeneous s-norm of degree k, $k > 0$, so that*

$$S(\lambda x, \lambda y) = \lambda^k S(x, y) \tag{3.4.3}$$

for all x, y, and λ in I. Then $k = 1$ and $S = Max$.

Proof. The substitution $x = 1$, $y = 0$ into (3.4.3) yields $\lambda = \lambda^k$, whence $k = 1$; and setting $x = y = 1$ in (3.4.3) yields $S(\lambda, \lambda) = \lambda$, i.e., $S = \text{Max}$. □

Homogeneous associative functions on infinite intervals play an important role in several studies concerning products of metric and normed spaces. The following theorem is due to F. Bohnenblust [1940]. The machinery we have developed allows us to give a simple proof.

Theorem 3.4.5 *Let G be an element of $\mathcal{G}_{St}(\mathbb{R}^+)$. Then G is homogeneous of degree k, $k > 0$, i.e.,*

$$G(\lambda u, \lambda v) = \lambda^k G(u, v), \tag{3.4.4}$$

for all u, v, and λ in $\mathbb{R}^+$, if and only if $k = 1$ and there exists a constant $c > 0$ such that $G(u, v) = (u^c + v^c)^{1/c}$.

Proof. Setting $v = 0$ and $u = 1$ in (3.4.4) yields $k = 1$. Next, since G admits the representation $G(u, v) = g^{-1}(g(u) + g(v))$, where g is a continuous, strictly increasing function from $\mathbb{R}^+$ onto $\mathbb{R}^+$ with $g(0) = 0$, it follows from (3.4.4) that

$$\frac{1}{\lambda} g^{-1}\left(g(\lambda u) + g(\lambda v)\right) = g^{-1}(g(u) + g(v)).$$

For a fixed λ, this last equality shows that $g(u)$ and $g(\lambda u)$ both generate G, and it follows that there is a constant $k > 0$ such that $g(\lambda u) = kg(u)$. As λ varies this k will depend upon λ, and thus we have

$$g(\lambda u) = k(\lambda) g(u). \tag{3.4.5}$$

Setting $u = 1$ in (3.4.5) we obtain $g(\lambda) = k(\lambda)g(1)$ and (3.4.5) becomes the classical Cauchy equation $g(\lambda u) = g(\lambda)g(u)/g(1)$. So $g(u) = g(1)u^c$ for some $c > 0$. Thus $G(u, v) = (u^c + v^c)^{1/c}$, and the theorem is proved. □

Theorem 3.4.6 *Let H be a binary operation on $\mathbb{R}^+$ which has 1 (rather than 0) as identity and is homogeneous of degree k, for some $k > 0$, i.e., $H(\lambda u, \lambda v) = \lambda^k H(u, v)$, for all u, v, and λ in $\mathbb{R}^+$. Then $k = 2$ and $H(u, v) = uv$.*

Proof. Since 1 is an identity and H is homogeneous of degree k, we have $H(u, v) = H\left(u \cdot 1, u \cdot \frac{v}{u}\right) = u^k H\left(1, \frac{v}{u}\right) = u^{k-1} v$, and for $v = 1$ this yields $u = u^{k-1}$, i.e., $k = 2$, whence $H(u, v) = uv$. □

In questions dealing with the synthesis of judgements [Aczél and Alsina (1984)] as well as in problems concerning products of probabilistic metric spaces [Alsina (1978)], one encounters the functional equation

$$T(x^\lambda, y^\lambda) = T(x, y)^\lambda, \tag{3.4.6}$$

where $\lambda > 0$ is arbitrary and T is a continuous t-norm which is to be found. This equation was first solved in [Aczél and Alsina (1984)]. We can present a simpler proof based on Theorem 3.4.5.

Theorem 3.4.7 *A continuous t-norm T satisfies the functional equation (3.4.6) for all $\lambda > 0$ if and only if $T =$ Min or T belongs to Family (2.6.4), i.e., there exists a constant $c > 0$ such that*

$$T(x, y) = \exp\Big(-[(-\log x)^c + (-\log y)^c]^{1/c}\Big). \tag{3.4.7}$$

Proof. Suppose that T has an interior idempotent, say a, and consider any b in (0,1). Then, using (3.4.6), we have

$$\begin{aligned} T(b,b) &= T\left(a^{\log b/\log a}, a^{\log b/\log a}\right) = T(a,a)^{\log b/\log a} \\ &= a^{\log b/\log a} = b. \end{aligned}$$

Thus b is also idempotent and it follows that either $T =$ Min or that T is Archimedean.

Suppose next that T is Archimedean but not strict. Then there is an a in (0,1) such that $T(a, a) = 0$. Consider any b in (0,1). If $b \leq a$, then $T(b, b) = 0$. If $b > a$, then there exists an integer n such that $b^n < a$ and we have

$$0 \leq T(b, b)^n = T(b^n, b^n) \leq T(a, a) = 0,$$

so again $T(b, b) = 0$. Thus $T(b, b) = 0$ for all b in $[0, 1)$. But this contradicts the continuity of T, whence T is strict.

Now define $G : \mathbb{R}^+ \times \mathbb{R}^+ \to \mathbb{R}^+$ by $G(u, v) = -\log T(e^{-u}, e^{-v})$. Then G belongs to $\mathcal{G}_{st}(\mathbb{R}^+)$. Furthermore, by (3.4.6), G is homogeneous of degree 1. Thus, by Theorem 3.4.5, we have $G(u, v) = (u^c + v^c)^{1/c}$ for some $c > 0$, whence (3.4.7) follows. □

It is remarkable that the conclusion of Theorem 3.4.7 holds under a much weaker hypothesis. Specifically, we have

Theorem 3.4.8 *Suppose that T is a continuous t-norm that satisfies (3.4.6) for two fixed positive numbers p and q such that $\log p/\log q$ is irrational. Then $T =$ Min or T is given by (3.4.7), for some $c > 0$.*

Proof. Successively setting $\lambda = p$ and $\lambda = q$ in (3.4.6), it follows by induction that, for all non-negative integers m and n,

$$T\left(x^{p^m q^n}, y^{p^m q^n}\right) = (T(x, y))^{p^m q^n}; \tag{3.4.8}$$

and letting $u = x^{p^m q^n}$ and $v = y^{p^m q^n}$ it follows that (3.4.8) also holds for all non-positive integers m and n. Thus (3.4.8) holds for all m and n in

$\mathbb{Z}$. Since $\log p / \log q$ is irrational, the set $\{p^m q^n : m, n \text{ in } \mathbb{Z}\}$ is dense in $\mathbb{R}^+$ (see [Niven (1956), pp. 71-72]. It therefore follows from the continuity of T that (3.4.6) holds for all $\lambda > 0$, and an appeal to Theorem 3.4.7 then completes the proof. □

3.5 Distributivity

In this section we consider the distributivity equation

$$A(x, B(y, z)) = B(A(x, y), A(x, z))$$

when one or both of the functions A and B is associative. Some extensions of these elementary results can be found in [Mak and Sigmon (1988)]. We begin with t-norms and s-norms.

Theorem 3.5.1 *Let T and T' be t-norms, and let S and S' be s-norms. Then, for all x, y, z in I:*

(a) $T(x, S(y, z)) = S(T(x, y), T(x, z))$ if and only if $S = Max$;
(b) $T(x, T'(y, z)) = T'(T(x, y), T(x, z))$ if and only if $T' = Min$;
(c) $S(x, T(y, z)) = T(S(x, y), S(x, z))$ if and only if $T = Min$;
(d) $S(x, S'(y, z)) = S'(S(x, y), S(x, z))$ if and only if $S' = Max$.

Proof. Simply let $y = z = 1$. □

In particular, note that Min is the only autodistributive t-norm and Max the only autodistributive s-norm.

Next we assume that one of A, B is a quasi-arithmetic mean (see the last part of Section 2.8).

Theorem 3.5.2 *Let T be a t-norm, let S be an s-norm, and let M be the weighted quasi-arithmetic mean generated by f, i.e.,*

$$M(x, y) = f^{-1}(pf(x) + qf(y)), \tag{3.5.1}$$

where p, q are in $(0, 1)$ with $p + q = 1$ and f is a continuous, strictly monotonic function from I onto some interval J. Then for all x, y, z in I:

(a) $M(x, T(y, z)) = T(M(x, y), M(x, z))$ if and only if $T = Min$;
(b) $M(x, S(y, z)) = S(M(x, y), M(x, z))$ if and only if $S = Max$.

Proof. To establish (a) let $x = y = z$. Since $M(x, x) = x$, this yields $pf(T(x, x)) = pf(x)$, whence $T(x, x) = x$, and $T = \text{Min}$. The proof of (b) is similar. □

Theorem 3.5.3 *Let T be a strict t-norm and let*

$$M(x, y) = h^{-1}(ph(x) + qh(y)), \tag{3.5.2}$$

where p, q are in $(0, 1)$ with $p + q = 1$ and $h : I \to I$ is continuous and strictly increasing, with $h(0) = 0$ and $h(1) = 1$. Then for all x, y, z in I:

$$T\,(x, M(y, z)) = M(T(x, y), T(x, z)) \tag{3.5.3}$$

if and only if

$$T(x, y) = h^{-1}(h(x)h(y)). \tag{3.5.4}$$

Similarly, if S is a strict s-norm, then

$$S(x, M(y, z)) = M(S(x, y), S(x, z))$$

if and only if

$$S(x, y) = h^{-1}(h(x) + h(y) - h(x)h(y)).$$

Proof. Suppose that T and M satisfy (3.5.3). Fix x in (0,1), and let $\sigma = \sigma_x$ be the x-section of T. Clearly, σ is strictly increasing and continuous. Moreover, by (3.5.3),

$$\sigma\,(M(y, z)) = M(\sigma(y), \sigma(z)),$$

whence, setting $\varphi = h \circ \sigma \circ h^{-1}$ and using (3.5.2), we have

$$\varphi(pu + qv) = p\varphi(u) + q\varphi(v), \tag{3.5.5}$$

for all u, v in I. Since $\varphi : I \to I$ is continuous, it follows from Lemma 2.8.4 that $\varphi(u) = au + b$. Since $\varphi(0) = 0$ and $\varphi(1) = h(x)$, we have $b = 0$ and $a = h(x)$, whence $\varphi(h(y)) = h(x)h(y)$. Thus, for all y in I,

$$h^{-1}(h(x)h(y)) = h^{-1} \circ \varphi \circ h(y) = \sigma(y) = T(x, y),$$

from which, since x was chosen arbitrarily, (3.5.4) follows.

The remainder of the proof of the first part is a matter of simple calculations; and the proof of the second part is similar. □

Note, in particular that the t-norm P is the unique strict t-norm which is distributive with respect to the weighted root-mean power

$$M(x,y) = \left(px^k + (1-p)y^k\right)^{1/k}, \quad k > 0.$$

Now we turn our attention to functions defined on $\mathbb{R}^+$ (see [Aczél (1966), pp. 318-319]).

Theorem 3.5.4 *Let H and G be two binary operations on $\mathbb{R}^+$. Suppose that G belongs to $\mathcal{G}_{St}(\mathbb{R}^+)$. Then H is distributive with respect to G, i.e.,*

$$H(u, G(v,w)) = G(H(u,v), H(u,w)), \tag{3.5.6}$$

if and only if

$$H(u,v) = g^{-1}(c(u)g(v)),$$

where g is an additive generator of G and c is an arbitrary function.

The parallel result holds when G is replaced by F in $\mathcal{F}_{st}(\mathbb{R}^+)$.

Proof. From (3.5.6) and the representation of G, we have

$$H(u, g^{-1}(g(v)+g(w))) = g^{-1}(g(H(u,v)) + g(H(u,w))).$$

Let $g(v) = x$, $g(w) = y$ and $K(u,x) = g(H(u, g^{-1}(x)))$. Then

$$K(u, x+y) = K(u,x) + K(u,y);$$

since K is bounded from below, it follows that $K(u,x) = c(u)x$, whence $H(u,v) = g^{-1}(c(u)g(v))$. □

We conclude this section with a system of distributivity equations of interest in the context of fuzzy operations. The result extends the characterization given in [Bertoluzza (1977)].

Theorem 3.5.5 *Suppose that $S, T : I^2 \to I$ satisfy the following conditions:*

(i) *S is non-decreasing in each place;*
(ii) *$S(x,0) = S(0,x) = x$, for all x in I;*
(iii) *T is commutative;*
(iv) *$T(x,1) = x$, for all x in I;*
(v) *$T(x, S(y,z)) = S(T(x,y), T(x,z))$,*
and
$S(x, T(y,z)) = T(S(x,y), S(x,z))$, for all x, y, z in I.

Then $S = Max$ and $T = Min$.

Proof. Since $S(1,0) = 1$ and S is non-decreasing, $S(1,1) = 1$. Hence $S(x,x) = S(T(x,1), T(x,1)) = T(x, S(1,1)) = T(x,1) = x$, and it follows from (ii) that $S = \text{Max}$.

Now for $y \leq z$, $T(x,z) = T(x, \text{Max}(y,z)) = \text{Max}[T(x,y), T(x,z)]$, i.e., $T(x,y) \leq T(x,z)$, and so T is non-decreasing in each place. But then $T(0,0) \leq T(0,1) = 0$, so that $T(0,0) = 0$. Consequently, $T(x,x) = T[\text{Max}(x,0), \text{Max}(x,0)] = \text{Max}[x, T(0,0)] = \text{Max}(x,0) = x$, whence $T = \text{Min}$. □

Other related distributivity equations and generalizations can be found in [Aczél (1966)].

3.6 Conical t-norms

The graph of a continuous t-norm T is the surface over I^2 defined by $z = T(x,y)$. If T is a non-strict Archimedean t-norm, then $z = 0$ on $\mathcal{Z}(T)$, the zero set of T. Suppose that the remainder of the surface consists of the line segments joining the points on the boundary curve of $\mathcal{Z}(T)$ to the point (1,1,1). What can be said about T? The purpose of this section is to present the complete answer to this question ([Alsina and Sklar (1987); Alsina (1986a)]).

To be precise, given a strict involution g, we generate a portion of a cone with vertex (1,1,1) (see Figure 3.6.1) as follows: Let (x,y) be a point in I^2 other than (1,1). If $y \leq g(x)$, define $T(x,y) = 0$. If $y > g(x)$ then there is a unique x_0 in I such that the points $(x_0, g(x_0))$, (x,y), and (1,1) are collinear. In this case, define

$$T(x,y) = \begin{cases} \frac{x-x_0}{1-x_0}, & x_0 \neq 1, \\ y, & x_0 = 1. \end{cases} \tag{3.6.1}$$

A non-strict, continuous Archimedean t-norm constructed in this manner is said to be **conical**.

Theorem 3.6.1 *Let T be a continuous, non-strict Archimedean t-norm with additive generator t satisfying $t(0) = 1$. Then T is conical if and only if the function f defined on I by*

$$f(x) = 1 - t^{-1}(x) \tag{3.6.2}$$

is a solution of the functional equation

$$f^{-1}\left(\frac{f(x)}{f(x+y)}\right)+f^{-1}\left(\frac{f(y)}{f(x+y)}\right)=1, \tag{3.6.3}$$

on the restricted domain $\{(x,y)$ *in* $I^2 : 0 < x+y \le 1\}$.

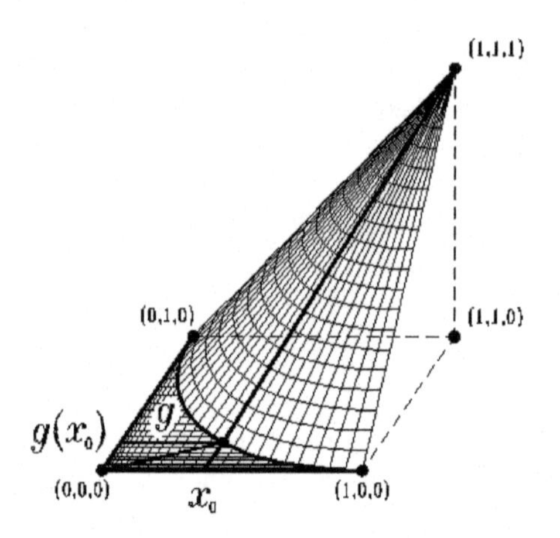

Figure 3.6.1. The construction (3.6.1).

Proof. Suppose that T is conical and let $g(x) = t^{-1}(1-t(x))$ be the equation of the boundary curve of $\mathcal{Z}(T)$. Then from (3.6.1) we have that, for any (x,y) in $I^2\backslash(1,1)$ such that $y > g(x)$,

$$x_0 = \frac{x-T(x,y)}{1-T(x,y)}, \quad \text{and} \quad g(x_0) = \frac{y-T(x,y)}{1-T(x,y)},$$

whence

$$g\left(\frac{x-T(x,y)}{1-T(x,y)}\right) = \frac{y-T(x,y)}{1-T(x,y)}.$$

Letting $u = t(x)$, $v = t(y)$, noting that $t(0) = 1$ and $t(1) = 0$, and using (3.6.2), it follows that

$$g\left(1-\frac{f(u)}{f(u+v)}\right) = 1-\frac{f(v)}{f(u+v)},$$

for all (u,v) in $I^2\backslash(1,1)$ such that $0 < u+v \le 1$; and from this, using the fact that

$$g(1-w) = 1 - f(1 - f^{-1}(w))$$

for all w in I, (3.6.3) readily follows.

Reversing the steps in the above argument yields the converse and completes the proof. □

Theorem 3.6.2 *Let f be a strictly increasing function from $\mathbb{R}^+$ to $\mathbb{R}^+$ with $f(0) = 0$. Furthermore, let f be such that both $f(x)/f(x+y)$ and $f(y)/f(x+y)$ are in Ran f whenever $x + y > 0$. Then f is a solution of the functional equation (3.6.3) if and only if there is a $c > 0$ such that, for all x in $\mathbb{R}^+$,*

$$f(x) = x^c. \tag{3.6.4}$$

Proof. Clearly, $f(x) = x^c$ with $c > 0$ satisfies (3.6.3) and the other conditions of the theorem. Conversely, let f satisfy (3.6.3) and these other conditions. Setting $y = x > 0$ in (3.6.3) yields

$$2f^{-1}\left(\frac{f(x)}{f(2x)}\right) = 1,$$

whence

$$f(x) = f\left(\frac{1}{2}\right) f(2x).$$

It follows that $f(1) = 1$ $\left(\text{set } x = \frac{1}{2}\right)$, and by induction that

$$f(2^n x) = (f(2))^n \cdot f(x)$$

for all integers n. Let $c = \log f(2)/\log 2 > 0$. Then $f(2) = 2^c$ and

$$f(2^n x) = 2^{cn} f(x) \tag{3.6.5}$$

for all integers n. In particular, $f(2^n) = 2^{cn}$ and $f^{-1}(2^{cn}) = 2^n$. Hence, upon setting $y = 3x$ in (3.6.3), we obtain

$$1 = \frac{1}{4} + f^{-1}\left(\frac{f(3x)}{4^c f(x)}\right),$$

whence

$$f(3x) = 4^c f\left(\frac{3}{4}\right) f(x) = f(3)f(x).$$

By induction, it follows that

$$f(3^n x) = f(3)^n f(x) \tag{3.6.6}$$

for all integers n.

Next, let $\{p_n/q_n\}$ be an increasing sequence of positive rational numbers and $\{r_n/s_n\}$ a decreasing sequence of positive rational numbers such that

$$\lim_{n\to\infty}(p_n/q_n) = \frac{\log 3}{\log 2} = \lim_{n\to\infty}(r_n/s_n).$$

The inequalities $(p_n/q_n) < \log 3/\log 2 < (r_n/s_n)$ yield

$$2^{p_n} < 3^{q_n} \text{ and } 3^{s_n} < 2^{r_n},$$

for all $n \geq 1$. Using (3.6.5), (3.6.6), and the fact that f is strictly increasing, we obtain

$$2^{cp_n} = f(2^{p_n}) < f(3^{q_n}) = f(3)^{q_n}$$

and

$$f(3)^{s_n} = f(3^{s_n}) < f(2^{r_n}) = 2^{cr_n}.$$

Thus

$$2^{c(p_n/q_n)} < f(3) < 2^{c(r_n/s_n)},$$

and so $f(3) = 2^{c(\log 3/\log 2)} = 3^c$, whence we immediately obtain:

$$f(2^n 3^m) = 2^{cn} f(3^m) = 2^{cn} 3^{cm} = (2^n 3^m)^c \tag{3.6.7}$$

for all integers n, m.

As a consequence of the well-known equidistribution theorem [Niven (1956), pp. 71-82], and the fact that log 3/log 2 is irrational, the set

$$\{2^n 3^m : n, m \text{ in } Z\}$$

is dense in $\mathbb{R}^+$. Thus from (3.6.7) and the strict monotonicity of f, we conclude that $f(x) = x^c$ for all x in $\mathbb{R}^+$. □

Corollary 3.6.3 *Let f be a strictly increasing function from I to I with $f(0) = 0$. Furthermore, let f be such that $f(x)/f(x+y)$ and $f(y)/f(x+y)$ are in Ran f for all x, y in I with $0 < x + y \leq 1$. Then f is a solution of (3.6.3) if and only if $f(x) = x^c$, with $c > 0$, holds for all x in I.*

Proof. Suppose that f satisfies (3.6.3) and define f^* on $\mathbb{R}^+$ by

$$f^*(x) = \begin{cases} f(x), & x \text{ in } I, \\ f(x/2^{n+1})/f(1/2)^{n+1}, & x \text{ in } [2^n, 2^{n+1}] \text{ for } n \geq 0. \end{cases}$$

It is readily verified that f^* is well-defined and satisfies the conditions of Theorem 3.6.3. Consequently, f^*, and hence f, satisfies (3.6.4). The converse is trivial. □

From the above results we obtain:

Theorem 3.6.4 *Let g be a strict involution. Then T is a conical t-norm if and only if there is a positive number c such that*

$$g(x) = 1 - [1 - (1 - x)^{1/c}]^c.$$

In this case, T is a member of Family (2.6.2), i.e., T is generated by $t(x) = (1 - x)^{1/c}$.

The preceding arguments, which date back to 1979, are of interest because they represent the first application of the Weyl equidistribution theorem in the theory of functional equations. They have subsequently been used in many other instances (see, e.g., Theorems 2.7.6 and 3.4.8). An alternate and simpler proof of the characterization of conical t-norms was subsequently given in [Alsina (1985a)]. It goes as follows:

Let T be a conical t-norm. For $0 < a < 1$, let g_a be the level curve at level a of T and let g_0 be the boundary curve of $\mathcal{Z}(T)$. Thus, for $0 \leq a < 1$,

$$g_a(x) = t^{-1}(t(a) - t(x)), \qquad \text{for } a \leq x \leq 1. \tag{3.6.8}$$

Since all the proper level curves of T are homothetic to the boundary curve of $\mathcal{Z}(T)$, for all a in [0,1) and all x in $[a, 1)$, we have

$$g_0\left(\frac{x - a}{1 - a}\right) = \frac{g_a(x) - a}{1 - a}.$$

Using (3.6.8), normalizing the generator t so that $t(0) = 1$, and letting $a = xy$, it follows that t satisfies the functional equation

$$t^{-1}\left(1 - t\left(\frac{x - xy}{1 - xy}\right)\right) = \frac{t^{-1}(t(xy) - t(x)) - xy}{1 - xy}, \tag{3.6.9}$$

for all x, y in I. Introducing the new variables $u = -\log(1 - xy)$, $v = -\log[(1 - x)/(1 - xy)]$ and the function $f : \mathbb{R}^+ \to I$ defined by

$f(x) = t(1 - e^{-x})$, a straightforward manipulation converts (3.6.9) to

$$f(u+v) + f\big(u + f^{-1}(1 - f(v))\big) = f(u),$$

for all u, v in $\mathbb{R}^+$. Thus for the t-norm T_f defined by

$$T_f(x,y) = f\big(f^{-1}(x) + f^{-1}(y)\big),$$

we have

$$T_f(x,y) + T_f(x, 1-y) = x.$$

By Lemma 3.2.12, $T_f(x,y) = xy$, and so $f^{-1}(x) = -c \log x$, for $c > 0$, i.e., $t(x) = (1-x)^{1/c}$.

Special classes of ruled surfaces are developable surfaces, which have Gaussian curvature zero. In this context we have the following

Theorem 3.6.5 *Let T be a non-strict Archimedean t-norm with additive generator t such that $t(0) = 1$. Suppose that t'' exists and is continuous on $(0,1)$ and that $t''(x) \neq 0$ for all x in $(0,1)$. Then the surface $z = T(x,y)$ is developable on $\{(x,y) \mid T(x,y) > 0\}$, i.e.,*

$$D_{11}T \cdot D_{22}T - (D_{12}T)^2 = 0, \ \textit{for all } (x,y) \textit{ in } (0,1)^2 \textit{ such that } T(x,y) > 0,$$

if and only if $t(x) = (1-x)^\alpha$ for some $\alpha > 0$.

Proof. Since $T(x,y) = t^{-1}\left(t(x) + t(y)\right)$ whenever $T(x,y) > 0$, we have

$$\begin{aligned}
D_{11}T(x,y) &= \frac{t''(x)}{t'(T(x,y))} - \frac{t'(x)^2 t''(T(x,y))}{t'(T(x,y))^3},\\
D_{22}T(x,y) &= \frac{t''(y)}{t'(T(x,y))} - \frac{t'(y)^2 t''(T(x,y))}{t'(T(x,y))^3},\\
D_{12}T(x,y) &= \frac{-t'(x)t'(y)t''(T(x,y))}{t'(T(x,y))^3}.
\end{aligned}$$

If the Gaussian curvature is zero, then

$$t''(x)t''(y)t'(T(x,y))^2 - t''(x)t'(y)^2 t''(T(x,y)) - t''(y)t'(x)^2 t''(T(x,y)) = 0,$$

or equivalently,

$$\frac{t'(T(x,y))^2}{t''(T(x,y))} = \frac{t'(y)^2}{t''(y)} + \frac{t'(x)^2}{t''(x)}.$$

Letting $u = t(x)$ and $v = t(y)$ we have

$$\frac{t'(t^{-1}(u+v))^2}{t''(t^{-1}(u+v))} = \frac{t'(t^{-1}(u))^2}{t''(t^{-1}(u))} + \frac{t'(t^{-1}(v))^2}{t''(t^{-1}(v))},$$

for all (u, v) in $(0,1)^2$ such that $u + v < 1$. Thus the function $g(u) = t'(t^{-1}(u))^2/t''(t^{-1}(u))$ satisfies the classical Cauchy equation $g(u+v) = g(u) + g(v)$ on the above-mentioned restricted domain. Since g is continuous, there exists a constant c such that $g(u) = cu$, i.e., $t'(t^{-1}(u))^2 = ct''(t^{-1}(u))u$ [Aczél (1966)]. Hence

$$t'(x)^2 = ct''(x)t(x),$$

for all x in (0,1). For $c = 0$ we obtain the contradiction $t'(x) = 0$. For $c = 1$ the general solution is $t(x) = e^{Ax+B}$ which cannot satisfy $t(1) = 0$. For $c \neq 0, 1$ elementary manipulations yield the desired result. The converse is immediate. □

3.7 Rational Archimedean t-norms

The principal aim of this section is to give a complete characterization of rational continuous Archimedean t-norms. This characterization is based on the following important

Theorem 3.7.1 *Let R be a (real or complex) rational two-place function which is reduced to lowest terms. Specifically, let*

$$R(x, y) = \frac{P(x, y)}{Q(x, y)},$$

where P and Q are relatively prime polynomials, neither of which is identically zero. Then R is commutative and associative, i.e., both $R(x, y)$ and $R(R(x, y), z)$ are symmetric functions, if and only if

$$R(x, y) = \frac{a_1xy + b_1(x+y) + c_1}{a_2xy + b_2(x+y) + c_2}, \tag{3.7.1}$$

where

$$\begin{aligned} b_1b_2 &= a_2c_1, \\ b_1^2 + b_2c_1 &= a_1c_1 + b_1c_2, \\ b_2^2 + a_2b_1 &= a_1b_2 + a_2c_2. \end{aligned} \tag{3.7.2}$$

Proof. Suppose first that R is commutative and associative. Then, by commutativity, we have

$$P(x,y)Q(y,x) = P(y,x)Q(x,y).$$

Since $P(x,y)$ and $Q(x,y)$ are relatively prime, we have that $P(x,y)|P(y,x)$ and similarly that $P(y,x)|P(x,y)$. Thus $P(x,y) = \pm P(y,x)$, whence the degree of P in its first place is equal to the degree of P in its second place. We denote this common value by deg P and, in the sequel, let $m = \deg P$. Similarly, we let $n = \deg Q$.

Next, by associativity, we have

$$R(R(x,y),z) = \frac{P(P(x,y)/Q(x,y),z)}{Q(P(x,y)/Q(x,y),z)} = \frac{G(x,y,z)}{H(x,y,z)}, \tag{3.7.3}$$

where G and H are polynomials. Expansion of the middle term in (3.7.3) yields that G and H are relatively prime and of degrees m and n, respectively, in z; and the associativity of R yields that m and n are also the degrees of G and H, respectively, in both x and y. Thus we may unambiguously write

$$\deg G = m, \quad \deg H = n.$$

Now fix a real number z_0 such that the set of roots $\{a_1, a_2, ..., a_m\}$ of $P(t,z_0) = 0$ and the set of roots $\{b_1, b_2, ..., b_n\}$ of $Q(t,z_0) = 0$ are disjoint. (The existence of z_0 is an easy consequence of the fact that we have $P(x,y) = 0 = Q(x,y)$ for at most finitely many points (x,y); see [Bôcher (1907), p. 210] for a proof via an extension of the Euclidean algorithm. Here we choose a z_0 different from the y-coordinates of these points.)

Let $M(x,y) = G(x,y,z_0)$ and $N(x,y) = H(x,y,z_0)$. Then M and N are symmetric with $\deg M = m$, and $\deg N = n$. From (3.7.3), abbreviating $P(x,y)$ to P and $Q(x,y)$ to Q, we have

$$\begin{aligned}\frac{M(x,y)}{N(x,y)} = \frac{P(P/Q,z_0)}{Q(P/Q,z_0)} &= k \cdot \frac{\prod_{i=1}^{m} [P/Q - a_i]}{\prod_{j=1}^{n} [P/Q - b_j]} \\ &= kQ^{n-m} \cdot \frac{\prod_{i=1}^{m} (P - a_iQ)}{\prod_{j=1}^{n} (P - b_jQ)}.\end{aligned} \tag{3.7.4}$$

Clearly, for any c, the polynomials Q and $P - cQ$ have no common non-constant factor; and for $a \neq b$, the same is true of $P - aQ$ and $P - bQ$: for if $P - aQ = FA$ and $P - bQ = FB$, then

$$(b-a)P = F(bA - aB) \text{ and } (b-a)Q = F(A-B),$$

which implies that $F|(b-a)P$ and $F|(b-a)Q$, i.e., that F is constant. Thus the last expression in (3.7.4) is in lowest terms. This observation allows us to determine the possible values of $m = \deg P$ and $n = \deg Q$. The argument splits into three cases:

1. $m < n$. If some b_j is zero, then all a_i are non-zero and $\deg(P - a_iQ) = n$, whence

$$m = \deg G = \deg M = (n-m)n + mn = n^2 \geq n,$$

which is a contradiction. Thus each b_j is non-zero so that $n = \deg H = \deg N = n^2$, i.e., $n = 0$ or 1, whence $m = 0$ and $n = 1$.

2. $m > n$. Arguing as in Case 1 (with N/M instead of M/N), we have $m = \deg M = m^2$, from which it follows that $m = 1$ and $n = 0$.

3. $m = n$. Since $\deg(P - cQ) = m$ for all but at most one value of c, it follows that either $m = \deg M = m^2$ or $n = \deg N = n^2$, whence m and n are either both 0 or both 1.

In all cases, therefore, P and Q are constant or bilinear. The commutativity of R yields that R is of the form (3.7.1). And, finally, a straightforward calculation shows that if R, as given by (3.7.1), is associative then the coefficients in (3.7.1) must satisfy (3.7.2).

Conversely, if R is given by (3.7.1) and the coefficients satisfy (3.7.2), then R is obviously commutative and we find that

$$R(R(x,y),z) = \frac{Axyz + B(xy + xz + yz) + C(x + y + z) + D}{Exyz + F(xy + xz + yz) + C(x + y + z) + D},$$

whence R is also associative. □

Note: The proof of Theorem 3.7.1 is based on arguments given in [Alt (1940)], modified and amended as necessary (see also [Fenyö and Paganoni (1987); Marley (1982)]). The observation that an expression of the form (3.7.1) is associative if and only if the coefficients satisfy (3.7.2) may be found in [Dickson (1916)].

As an immediate consequence of Theorem 3.7.1, we have the following result which is due to E.T. Bell [1936].

Corollary 3.7.2 *A two-place polynomial function P is commutative and associative if and only if*

$$P(x,y) = axy + b(x+y) + c, \tag{3.7.5}$$

where $b^2 = ac + b$.

Now suppose that T satisfies (3.7.1) and (3.7.2) and the boundary condition $T(x,1) = x$ for all x in I. Then

$$(a_2 + b_2)x^2 + (b_2 + c_2 - a_1 - b_1)x - (b_1 + c_1) = 0,$$

for all x (or for all x in I). Thus, in addition to (3.7.2), the coefficients in (3.7.1) must satisfy

$$a_2 + b_2 = b_1 + c_1 = 0 \text{ and } a_1 + b_1 = b_2 + c_2.$$

A simple computation now yields that

$$T(x,y) = \frac{(a_1+b_1)xy - b_1(1-x)(1-y)}{(b_2+c_2) - b_2(1-x)(1-y)}.$$

If $a_1 + b_1 = b_2 + c_2 = 0$, then T is constant, contradicting $T(x,1) = x$. Consequently, letting $\beta = b_1/(a_1+b_1)$ and $\alpha = b_2/(b_2+c_2)$, it follows that any commutative and associative rational function T that satisfies the boundary condition $T(x,1) = x$ depends only on the two parameters α and β and must be of the form

$$T_{\alpha,\beta}(x,y) = \frac{xy - \beta(1-x)(1-y)}{1 - \alpha(1-x)(1-y)}. \tag{3.7.6}$$

Furthermore, $T_{\alpha,\beta}$ satisfies the boundary condition $T_{\alpha,\beta}(x,0) = 0$ for all x (or for all x in I) if and only if $\beta = 0$.

Next, $T_{\alpha,0}$ is continuous on I^2 if and only if $\alpha \le 1$. Since $T_{\alpha,0}$ is increasing on $(0,1]^2$ when $\alpha \le 1$, this yields

Corollary 3.7.3 *A rational function (restricted to I^2) is a t-norm if and only if it is of the form $T_{\alpha,0}$ with $\alpha \le 1$. In this case it is a strict t-norm – specifically, a member of Family (2.6.3).*

We now turn to the consideration of non-strict Archimedean t-norms that are rational on the complement of their zero set. For any such t-norm $T_{\alpha,\beta}$, the boundary of $\mathcal{Z}(T_{\alpha,\beta})$ is the portion of the curve

$$xy - \beta(1-x)(1-y) = 0 \tag{3.7.7}$$

which lies in I^2. The existence of such a portion necessitates that $\beta > 0$. For $\beta = 1$, this portion is the line segment joining the points (0,1) and (1,0). For $\beta \neq 1$, it is the segment of a hyperbola which passes through the points $(0,1)$ and $(1,0)$, has asymptotes $x = \beta/(\beta-1)$ and $y = \beta/(\beta-1)$ and intersects the diagonal of I^2 at the point (γ,γ), where $\gamma = \sqrt{\beta}/(1+\sqrt{\beta})$. For $\beta < 1$, this segment is convex; for $\beta > 1$, it is concave.

It is easy to show that the maximum value of $(1-x)(1-y)$ on the boundary of $\mathcal{Z}(T_{\alpha,\beta})$ is attained at the point (γ,γ) and equals $1/(1+\sqrt{\beta})^2$. Thus, if $\alpha < (1+\sqrt{\beta})^2$ then $T_{\alpha,\beta}$ is continuous on the open set $(0,1)^2\backslash\mathcal{Z}(T_{\alpha,\beta})$. Furthermore, $T_{\alpha,\beta}$ is increasing on this set if and only if

$$D_2T_{\alpha,\beta}(x,y) = \frac{\alpha x^2 + (1-\alpha-\beta)x + \beta}{[1-\alpha(1-x)(1-y)]^2} > 0$$

for all x, y in $(0,1)^2\backslash\mathcal{Z}(T_{\alpha,\beta})$, i.e., if and only if

$$p(x) = \alpha x^2 + (1-\alpha-\beta)x + \beta > 0, \text{ for all } x \text{ in } (0,1), \tag{3.7.8}$$

where, as previously noted, $\beta > 0$.

If $\alpha = 0$, then $p(x) = x + \beta(1-x)$, whence (3.7.8) holds. In this case, $T_{0,\beta}$ is a member of Family (2.6.7).

For $\alpha \neq 0$, the graph of p is a parabola which passes through the points $(0,\beta)$ and (1,1) – both in the upper half-plane – and whose vertex is the point $(v, p(v))$, where

$$v = \frac{\alpha+\beta-1}{2\alpha}, \quad p(v) = -\Delta(\alpha,\beta)/4\alpha,$$

and

$$\Delta(\alpha,\beta) = (1-\alpha-\beta)^2 - 4\alpha\beta.$$

If $\alpha < 0$, this parabola is directed downward; in this case, its vertex must also lie in the upper half-plane and it follows that $p(x) > 0$ for all x in I. When $\alpha > 0$, the parabola is directed upward. In this case (3.7.8) holds unless $0 < v < 1$ and $p(v) \leq 0$, or equivalently, provided that one of the following conditions is satisfied:

(a) $p(v) > 0$, i.e., $\Delta(\alpha, \beta) < 0$;
(b) $v \le 0$, i.e., $\alpha + \beta \le 1$;
(c) $v \ge 1$, i.e., $\beta - \alpha \ge 1$.

Combining these observations and noting that, in the $\alpha - \beta$ plane, the graph of $\sqrt{\alpha} = 1 + \sqrt{\beta}$ is that portion of the parabola $\Delta(\alpha, \beta) = 0$ for which $1 \le \alpha \le \beta$, yields:

Theorem 3.7.4 *The function $T_{\alpha,\beta}$ defined on I^2 by*

$$T_{\alpha,\beta}(x, y) = \frac{\max[xy - \beta(1-x)(1-y), 0]}{1 - \alpha(1-x)(1-y)} \tag{3.7.9}$$

is a continuous Archimedean t-norm if and only if $\beta \ge 0$ and either (i) $\alpha \le 1$ or (ii) $\sqrt{\alpha} < 1 + \sqrt{\beta}$, i.e., if and only if the point (α, β) lies in the shaded region in Figure 3.7.1. Furthermore, the family $\{T_{\alpha,\beta}\}$ is positively directed in α and negatively directed in β.

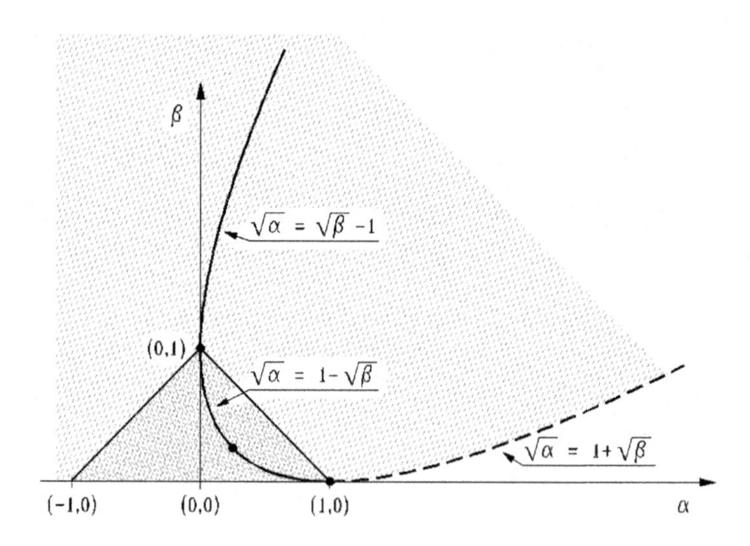

Figure 3.7.1. The parameter space in Theorem 3.7.4.

As previously mentioned, the subfamily $\{T_{\alpha,0}\}$ is the strict Family (2.6.3), and the subfamily $\{T_{0,\beta}\}$ is Family (2.6.7). Similarly, the subfamily $\{T_{(1-\sqrt{\beta})^2,\beta}\}$ is Family (2.6.8). Note that this subfamily is determined by those points (α, β) which, for $\beta > 0$, lie on the portion of the parabola

$\Delta(\alpha, \beta) = 0$ that is in the interior of the shaded region in Figure 3.7.1. Note also that

$$T_{0,0} = P, \quad T_{0,1} = W \quad \text{and} \quad T_{1,0} = P/(\Sigma - P).$$

It is also easily verified that

(i) $T_{\alpha,\beta} \le P$ if and only if $\alpha \le \beta$; and
(ii) $T_{\alpha,\beta} \ge W$ if and only if $\beta \le 1 + \min(\alpha, 0)$.

Now let $t_{\alpha,\beta}$ be an additive generator of $T_{\alpha,\beta}$. Since $T_{\alpha,\beta}$ is differentiable and both $D_1 T_{\alpha,\beta}$ and $D_2 T_{\alpha,\beta}$ are different from 0 on $(0,1)^2 \backslash \mathcal{Z}(T_{\alpha,\beta})$, it follows from the remarks after Corollary 2.2.14 that $t_{\alpha,\beta}$ is differentiable on (0,1). Consequently, (2.2.21) holds and we have

$$\frac{t'_{\alpha,\beta}(x)}{t'_{\alpha,\beta}(y)} = \frac{D_1 T_{\alpha,\beta}(x,y)}{D_2 T_{\alpha,\beta}(x,y)} = \frac{p(y)}{p(x)},$$

where p is the polynomial given in (3.7.8). Thus

$$t'_{\alpha,\beta}(x) = k/(\alpha x^2 + (1 - \alpha - \beta)x + \beta),$$

from which it follows at once that $t(x)$ must take one of the forms

$$\frac{p^0 x + q^0}{r^0 x + s^0}, \quad \arctan\left(\frac{p^+ x + q^+}{r^+ x + s^+}\right) \quad \text{or} \quad \log\left(\frac{p^- x + q^-}{r^- x - s^-}\right),$$

depending on whether the discriminant $\Delta(\alpha, \beta)$ is (as indicated by the superscripts) zero, positive or negative (and where the coefficients obviously depend on α and β).

Here it is interesting to note that the one-parameter subfamily $\{T_{\alpha,\alpha}\}$, for which $\Delta(\alpha, \beta) = 1 - 4\alpha$, is generated by

$$\log\left[\frac{2 - (1 - \sqrt{\Delta})(1 - x)}{2 - (1 + \sqrt{\Delta})(1 - x)}\right], \text{ when } 0 \le \alpha < 1/4,$$

$$\frac{1 - x}{1 + x}, \qquad \text{when } \alpha = 1/4,$$

$$\arctan\left(\sqrt{-\Delta}\frac{1 - x}{1 + x}\right), \qquad \text{when } 1/4 < \alpha.$$

Next we have that

$$t''_{\alpha,\beta}(x) = -k(2\alpha x + (1 - \alpha - \beta))/p^2(x).$$

Since $p(x) > 0$ and $t'_{\alpha,\beta}(x) < 0$ on (0,1), we must have $k < 0$. Consequently, $t''_{\alpha,\beta}(x) > 0$ on (0,1) if and only if $1 - \alpha - \beta \geq 0$ and $1 + \alpha - \beta \geq 0$. This yields:

Theorem 3.7.5 *The function $T_{\alpha,\beta}$ defined by (3.7.9) is a copula if and only if $0 \leq \beta \leq 1 - |\alpha|$, i.e., if and only if the point (α, β) belongs to the darkly shaded triangle in Figure 3.7.1.*

Note, in particular, that $T_{\alpha,\alpha}$ is a copula if and only if $0 \leq \alpha \leq 1/2$. Note further that by immediately imposing the condition $t''_{\alpha,\beta} > 0$ one can go directly from (3.7.6) to Theorem 3.7.5 (see [Nelsen (1999)]).

In the preceding discussion we have assumed that $\alpha < (1 + \sqrt{\beta})^2$. But a very interesting situation arises when equality holds, i.e., when the point (α, β) is on the parabolic segment which bounds the parameter space in the first quadrant of the (α, β)-plane (see Figure 3.7.1). When $\beta = 0$, we have $\alpha = 1$ and, as observed previously, $T_{1,0} = P/(\Sigma - P)$. When $\beta > 0$, the function $T_{\alpha,\beta}$ given in (3.7.9) is undefined at the point (γ, γ), where again, $\gamma = \sqrt{\beta}/(1 + \sqrt{\beta})$. But since, for $a < \alpha = (1 + \sqrt{\beta})^2$, the function $T_{a,\beta}$ is a well-defined continuous t-norm, and since the family $\{T_{a,\beta}\}$ is positively directed in a, it follows from Theorem 2.5.5 that $\lim_{a \nearrow \alpha} T_{a,\beta}$ is a left-continuous t-norm. Thus we have:

Example 3.7.6 For any $\beta \geq 0$, and with $\gamma = \sqrt{\beta}/(1 + \sqrt{\beta})$, the function T_β defined on I^2 by

$$T_\beta(x, y) = \begin{cases} \dfrac{\max[xy - \beta(1-x)(1-y), 0]}{1 - (1 + \sqrt{\beta})^2(1-x)(1-y)}, & (x, y) \neq (\gamma, \gamma), \\ 0, & (x, y) = (\gamma, \gamma), \end{cases}$$

is a left-continuous t-norm. The t-norm T_β is constant and equal to γ on the segments $\{\gamma\} \times (\gamma, 1]$ and $(\gamma, 1] \times \{\gamma\}$, since – see (3.7.8) – $p(\gamma) = 0$. Elsewhere in $(0,1)^2 \backslash \mathcal{Z}(T_\beta)$, T_β is strictly increasing. Finally, T_β is continuous everyhere except at the point (γ, γ), where it jumps from 0 to γ (see Figure 3.7.2). Note that if, instead of $T_\beta(\gamma, \gamma) = 0$, we define $T_\beta(\gamma, \gamma) = \gamma$, then the resulting function is a t-norm which has an interior idempotent.

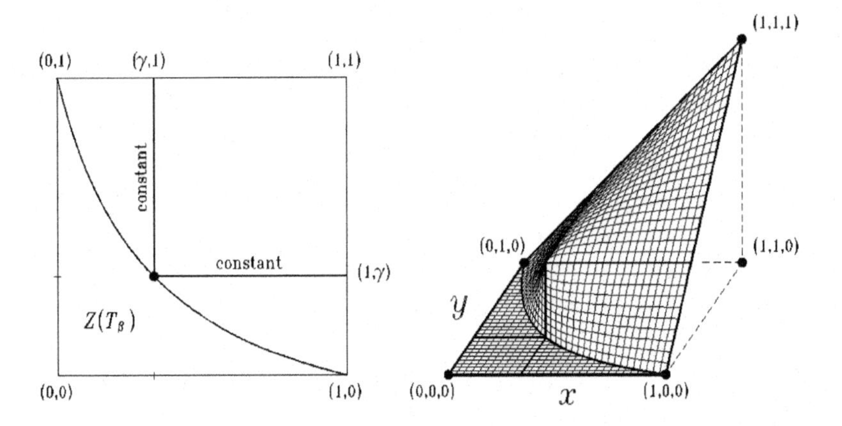

Figure 3.7.2. The t-norm $T_{1/2}$ in Example 3.7.6.

3.8 Extension and sets of uniqueness

In this section we investigate questions of existence and uniqueness of continuous Archimedean t-norms, given their restrictions to certain subsets of I^2. Questions of this nature were first considered by C. Kimberling [1973a].

More precisely, let E be a subset of I^2 and suppose that T_E is a function from E to I which is consistent with the conditions imposed on elements of $\mathcal{T}_{Ar}$. The problem is to determine whether T_E can be extended to a t-norm T in $\mathcal{T}_{Ar}$ and, if so, whether the extension is unique. The subset E will usually be a one-dimensional set consisting of line segments and/or simple curves. We shall obtain conditions under which T_E can be extended from E to I^2 and establish uniqueness of these extensions in a number of cases. The results will point out in a dramatic way the strength of the associativity condition.

In view of the Representation Theorem 2.1.6, these questions reduce to the study of existence and uniqueness of additive generators. Specifically, extending T_E to a t-norm T in $\mathcal{T}_{Ar}$ is equivalent to finding a generator of T, i.e., a decreasing function $t : I \to \mathbb{R}^+$ such that

$$t^{(-1)}(t(x) + t(y)) = T_E(x, y) \quad \text{for all } (x, y) \text{ in } E. \tag{3.8.1}$$

Moreover, by part (a) of Corollary 2.2.6, such an extension is unique if and only if any two solutions of (3.8.1) differ by a multiplicative constant. Note that it follows from (2.1.3) that (3.8.1) is equivalent to

$$t \circ T_E(x, y) = \min[t(x) + t(y), t(0)], \quad \text{for all } (x, y) \text{ in } E. \tag{3.8.2}$$

Our choices of E will lead to the consideration of several functional equations – in particular, to the classical equations of Cauchy, Schröder, and Abel mentioned in Section 1.2 – and to a number of interesting systems of such equations. We first consider the case when the given subset of I^2 is the set $D = \{(x, x) : x \text{ in } I\}$.

Definition 3.8.1 **A diagonal** is a function δ from I onto I which satisfies the following conditions:

(a) There is a number d in $[0, 1)$ such that $\delta(x) = 0$ for all x in $[0, d]$ and δ is strictly increasing on $[d, 1]$;
(b) $\delta(x) < x$ for all x in $(0, 1)$.

Note that δ is continuous on I.

In view of the results of Section 2.1, if T is in $\mathcal{T}_{Ar}$, then δ_T, the diagonal of T, satisfies the conditions (a) and (b) above with $d = t^{(-1)}\left(\frac{1}{2}t(0)\right)$ and t any generator of T. In particular, $d = 0$ if and only if T is strict.

In the other direction, suppose that δ is a diagonal and $T_D = \delta$. Then, with $E = D$, (3.8.2) yields the Schröder equation

$$t(\delta(x)) = 2t(x), \quad \text{for } d \leq x \leq 1. \tag{3.8.3}$$

Therefore, $\delta = \delta_T$ for some T in $\mathcal{T}_{Ar}$ if and only if every additive generator t of T is a solution of (3.8.3), and the problem of extension reduces to the construction of suitable solutions of this equation.

It is a straightforward matter to adapt the standard construction of the general solution given in [Kuczma (1968)] to fit the current conditions. Fix x_0 as follows: if $d = 0$, choose x_0 arbitrarily in (0,1); if $d > 0$, let x_0 be the unique point in $(d, 1)$ for which $\delta(x_0) = d$. Now let $x_n = \delta^n(x_0)$ for all n in $\mathbb{Z}$. If $d = 0$, we have $x_n < x_{n-1}$ for all n in $\mathbb{Z}$, whence $\lim_{n\to-\infty} x_n = 1$ and $\lim_{n\to\infty} x_n = 0$. If $d > 0$, we still have $\lim_{n\to-\infty} x_n = 1$, but now $x_n = 0$ for all $n \geq 1$. Next, let $\bar{t}$ be any strictly decreasing function from $[x_1, x_0]$ onto $[1, 2]$. Extend $\bar{t}$ to a function t on I by defining

$$t(x) = 2^n \bar{t}(\delta^{-n}(x)), \quad \text{for } x_{n+1} \leq x < x_n, \tag{3.8.4}$$

$t(1) = 0$, and $t(0) = \infty$ in case $d = 0$. Then t is a strictly decreasing and continuous solution of (3.8.3) with $t(x_0) = 1$, and every such solution arises in this way (see Figure 3.8.1).

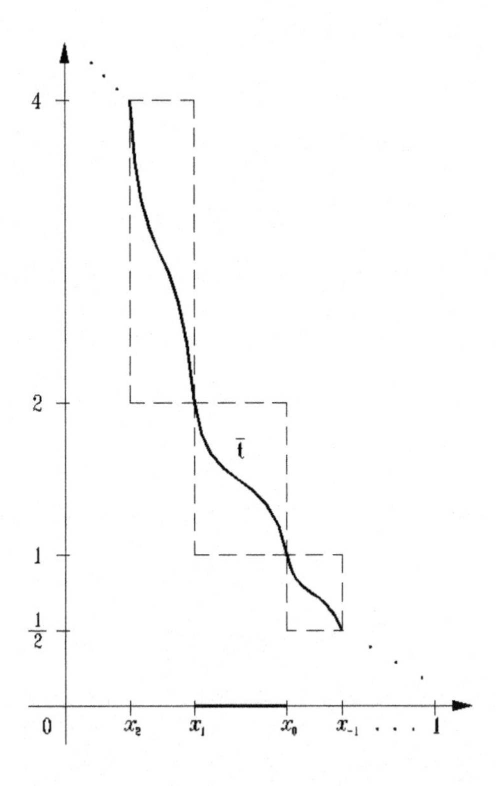

Figure 3.8.1. Solutions of (3.8.3).

Noting that each choice of $\bar{t}$ yields a solution t which generates a distinct T, we have proved:

Theorem 3.8.2 *For each diagonal δ, there are infinitely many distinct T in $\mathcal{T}_{Ar}$ for which $\delta_T = \delta$. The additive generators t of all such T depend on an arbitrary function $\bar{t}$ and are given by (3.8.4).*

The Schröder equation (3.8.3) may have a unique solution when further restrictions are imposed on δ or t. For instance, if T is a copula whose diagonal δ satisfies $\delta'(1-) = 2$, then (3.8.3) has a unique *convex* solution (up to constant multiples).

This fact can be gleaned from a standard result on Schröder's equation. Define γ on I via $\gamma(x) = 1-\delta^{-1}(1-x)$ for x in [0,1) and $\gamma(1) = \lim_{x \nearrow 1} \gamma(x)$. Then γ is continuous and strictly increasing, $\gamma(0) = 0$, and $\gamma(x) < x$ for all x in (0,1), and (3.8.3) is equivalent to

$$s(\gamma(x)) = \frac{1}{2}s(x), \text{ for all } x \text{ in } I,$$

where $s(x) = t(1-x)$; moreover, $\gamma'(0+) = 1/\delta'(1-) = 1/2$. We now invoke Theorem 6.6 of [Kuczma (1968)]: a convex solution s of this equation – hence a convex solution t of (3.8.3) – is unique (up to constant multiples).

Thus any copula satisfying $\delta'(1-) = 2$ is uniquely determined by its diagonal. This result, which was announced in [Frank (1996)], admits a significant and startling stochastic interpretation:

Suppose that the random variables X and Y are connected by the copula C, i.e., $C_{XY} = C$. The probability transforms $U = F_X(X)$ and $V = F_Y(Y)$ are uniformly distributed on I and $C_{UV} = C$ (see Theorem 1.4.4). Now for the random variable $M = \text{Max}(U, V)$ we have

$$F_M(x) = \Pr(\text{Max}(U, V) \leq x) = \Pr(U \leq x, V \leq x) = C(x, x),$$

in other words,

$$F_M(x) = \delta_C(x), \text{ for all } x \text{ in } I.$$

Thus if one assumes that the copula of two continuous random variables is Archimedean – a convenient assumption in many modeling situations – and if $F'_M(1-) = 2$, then their *joint* distribution is uniquely determined by the *one*-dimensional distribution F_M. For example, if $F_M(x) = x^2$ then, necessarily, $C = P$ so that X and Y must be independent.

The restriction $F'_M(1-) = 2$ is not nearly as stringent as it might seem at first sight. Thus, since $\delta'_W(1-) = \delta'_P(1-) = 2$, we have $\delta'_C(1-) = 2$ whenever $W \leq C \leq P$. This condition is satisfied by many other copulas as well: among the copulas in Table 2.6, by all members of Families 1, 3, 5, 8, 15, 18, 20, 22 and 23; and by all members of the two-parameter family in Example 2.6.2.

On the other hand, when $\delta'(1-) < 2$ or does not exist, equation (3.8.3) may (or may not) have distinct convex solutions. Among those that do are the diagonals δ_α of several families $\{T_\alpha\}$ in Table 2.6: Families 2, 4 and 6 for all $\alpha \neq 1$ $(\delta'_\alpha(1-) = 2^{1/\alpha})$ and Family 21 for all α $(\delta'_\alpha(1-) = 1)$. These diagonals can thus be extended to distinct copulas.

To illustrate the construction, let C_0 be the copula from Family (2.6.4) with $\alpha = 2$, so that

$$t_0(x) = \log^2 x \text{ and } \delta_0(x) = x^{\sqrt{2}}.$$

For β in (0,1], define $\varphi_\beta : \mathbb{R}^+ \to \mathbb{R}^+$ by

$$\varphi_\beta(u) = u + 2^n \beta \sin(u/2^n), \text{ when } 2^{n+1}\pi \le u \le 2^{n+2}\pi,$$

for all n in $\mathbb{Z}$, and $\varphi(0) = 0$. Clearly, φ_β is a continuous, strictly increasing, and non-trivial solution of $\varphi(2u) = 2\varphi(u)$. Now let $t_\beta = \varphi_\beta \circ t_0$. A tedious calculation shows that t_β is convex whenever $0 < \beta \le 1/(1+8\pi)$. For any such β, let C_β be the copula generated by t_β and δ_β its diagonal. Then

$$\begin{aligned} \delta_\beta(x) &= t_\beta^{-1}(2t_\beta(x)) = t_0^{-1} \circ \varphi_\beta^{-1}(2\varphi_\beta \circ t_0(x)) \\ &= t_0^{-1} \circ \varphi_\beta^{-1} \circ \varphi_\beta(2t_0(x)) = \delta_0(x), \end{aligned}$$

i.e., $\delta_\beta = \delta_0$. But for any distinct β, γ as above, $C_\beta \ne C_\gamma$.

The preceding discussion leads to an open problem: characterize those diagonals that can be extended to associative copulas, i.e., obtain necessary and sufficient conditions on δ under which (3.8.3) has a convex solution t.

Definition 3.8.3 For y in $(0,1)$, a **y-section** is a function σ_y from I onto $[0,y]$ which satisfies the following conditions:

(a) There is a number c in $[0,1)$ such that $\sigma_y(x) = 0$ for all x in $[0,c]$ and σ_y is strictly increasing on $[c,1]$;
(b) $\sigma_y(x) < x$ for all x in $(0,1]$.

If T is in $\mathcal{T}_{Ar}$, then any y-section of T satisfies the conditions (a) and (b) above with $c = t^{(-1)}[t(0) - t(y)]$ and t any generator of T, whence $c = 0$ if and only if T is strict.

The problem of extension of sections leads to an Abel equation. For a fixed y_0 in (0,1) and a given y_0-section σ, let $T_H = \sigma$ and $H = H_{y_0} = \{(x, y_0) : x \text{ in } I\}$. Since a generator is determined up to a positive multiple, we may, without loss of generality, choose t so that $t(y_0) = 1$. Then (3.8.2) reduces to

$$t(\sigma(x)) = t(x) + 1, \quad \text{for } c \le x \le 1, \tag{3.8.5}$$

where $c = t^{(-1)}[t(0) - 1]$ and $t(y_0) = 1$. Thus σ can be extended to some T in $\mathcal{T}_{Ar}$ so that σ is the y-section of T precisely when an additive generator t of T is a solution of (3.8.5).

As in the preceding case, we give the standard construction of the general solution, which again depends on an arbitrary function. For $n \geq -1$, let $y_n = \sigma^n(y_0)$. When $c = 0$, we have $y_n < y_{n-1}$ for all $n \geq 0$ and $\lim_{n\to\infty} y_n = 0$; and when $c > 0$, there is a least integer N such that $y_n = 0$ for all $n \geq N$. Now let $\bar{t}$ be any strictly decreasing function from $[y_0, 1]$ onto I. Extend $\bar{t}$ to a function t on (0,1] by defining

$$t(x) = n + \bar{t}(\sigma^{-n}(x)), \qquad \text{for } y_n < x \leq y_{n-1}, \tag{3.8.6}$$

for all $n \geq 0$ if $c = 0$, and for $n = 0, 1, ..., N$ if $c > 0$ (in which case $t(0) = N + 1$). This yields:

Theorem 3.8.4 *For each y-section σ, $0 < y < 1$, there are infinitely many distinct T in $\mathcal{T}_{Ar}$ whose y-section is σ. The additive generators t of these T for which $t(y) = 1$ depend on an arbitary function $\bar{t}$ and are given by (3.8.6).*

Note that the commutativity of T yields identical results for x-sections.

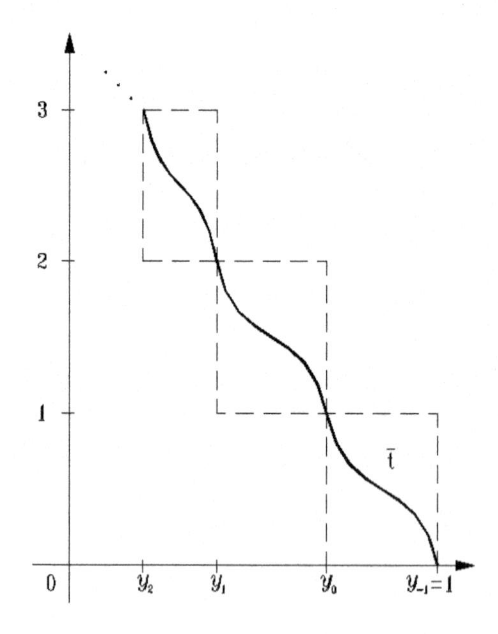

Figure 3.8.2. Solutions of (3.8.5).

For z in [0,1), a **z-level curve** $L(z)$ is the graph of any strictly decreasing function l from $[z, 1]$ to $[z, 1]$ such that $l^{-1} = l$. For a given T in $\mathcal{T}_{Ar}$ and z in (0,1), the level set

$$L_T(z) = \{(x, y) : T(x, y) = z\}$$

is a z-level curve. Moreover, when T is non-strict, the boundary of the zero set $\mathcal{Z}(T)$ is a 0-level curve.

The problem of extending functions that are constant on level curves to t-norms is readily solved. Fix z_0 in [0,1) and let

$$E = L(z_0) = \{(x, l(x)) : z_0 \leq x \leq 1\},$$

where l is as above, and let $T_E(x, l(x)) = z_0$, for $z_0 \leq x \leq 1$. Choose t so that $t(z_0) = 1$ (and note that if $z_0 = 0$ then T cannot be strict and $L(z_0)$ must be the boundary of $\mathcal{Z}(T)$). Then (3.8.2) becomes

$$t(x) + t(l(x)) = 1, \text{ for } z_0 \leq x \leq 1. \tag{3.8.7}$$

The general solution of this equation is easily obtained. Let x_0 be the fixed point of l, so that $l(x_0) = x_0$, and let $\bar{t}$ be any strictly decreasing function from $[x_0, 1]$ to $[0, \frac{1}{2}]$. Extend $\bar{t}$ to a function t on $[z_0, 1]$ via

$$t(x) = 1 - \bar{t}(l(x)), \text{ for } z_0 \leq x \leq x_0. \tag{3.8.8}$$

Since $l^2(x) = x$, the function t satisfies (3.8.7). If $z_0 = 0$, we are done. If $z_0 > 0$ then any appropriate (i.e., continuous and strictly decreasing) extension of t to I yields a generator having the desired properties. Thus there are many t-norms, strict and non-strict, which possess a specified z-level curve. This proves:

Theorem 3.8.5 *For fixed z in $(0, 1)$, each z-level curve is the level set $L_T(z)$ of infinitely many strict and non-strict T in $\mathcal{T}_{Ar}$. Each 0-level curve is the boundary of $\mathcal{Z}(T)$ for infinitely many T in $\mathcal{T}_{Ar}$. The additive generators of all such T depend on an arbitrary function.*

To illustrate, the quarter disc $\{(x, y) \text{ in } I^2 : x^2 + y^2 \leq 1\}$ is the common zero set of the t-norms

$$T_1(x, y) = \max\left[(x^2 + y^2 - 1)^{1/2}, 0\right],$$
$$T_2(x, y) = \max\left[xy - ((1 - x^2)(1 - y^2))^{1/2}, 0\right],$$

whose generators are $t_1(x) = 1 - x^2$ and $t_2(x) = 1 - \frac{2}{\pi} \arcsin x$, respectively.

We now turn from questions of existence to questions of uniqueness and non-uniqueness.

Definition 3.8.6 A subset U of I^2 is a **set of uniqueness** for $\mathcal{T}_{Ar}$ if each T_0 in $\mathcal{T}_{Ar}$ is uniquely determined by its restriction to U, i.e., if for all T in $\mathcal{T}_{Ar}$,

$$T = T_0 \text{ on } U \text{ implies } T = T_0 \text{ on } I^2. \tag{3.8.9}$$

The preceding discussion shows that the diagonal of I^2, sections of I^2 and level curves in I^2 are not sets of uniqueness. There are also two-dimensional sets of non-uniqueness.

Theorem 3.8.7 *For each $\epsilon > 0$, the square $[0, 1-\epsilon]^2$ is not a set of uniqueness for $\mathcal{T}_{Ar}$.*

Proof. Consider T_1, T_2 in $\mathcal{T}_{Ar}$ with respective generators t_1, t_2 and suppose that

$$t_1(x) = t_2(x), \text{ for } 0 \le x \le 1-\epsilon,$$

and

$$t_1(x) < t_2(x), \text{ for } 1-\epsilon < x < 1.$$

Clearly $T_1(x,y) = T_2(x,y)$ for all (x,y) in $[0,1-\epsilon]^2$. Next, let y_0 be in $(0, 1-\epsilon)$ and note that $t_1(0) > t_1(y_0) = t_2(y_0) > t_2(1-\epsilon)$. Now choose x_0 in $(1-\epsilon, 1)$ so that $t_1(x_0)+t_1(y_0) < t_1(0)$. Then, since $t_1(x_0) < t_2(x_0)$ and $t_2(x_0) + t_2(y_0) > t_2(1-\epsilon)$, we have

$$\begin{aligned} T_1(x_0, y_0) &= t_1^{(-1)}(t_1(x_0) + t_1(y_0)) \\ &> t_1^{(-1)}(t_2(x_0) + t_2(y_0)) \\ &= t_2^{(-1)}(t_2(x_0) + t_2(y_0)) = T_2(x_0, y_0), \end{aligned}$$

whence $T_1 \ne T_2$. □

It follows from Theorem 3.8.7 that if U is a set of uniqueness for $\mathcal{T}_{Ar}$ then, for *every* neighborhood V of the upper boundary of I^2, the intersection of U with $V \cap (0,1)^2$ must be *non-empty*.

The condition (3.8.9) translates into the following statement about the generators t and t_0 of T and T_0, respectively: If

$$t^{(-1)}(t(x) + t(y)) = t_0^{(-1)}(t_0(x) + t_0(y)), \quad \text{for all } (x,y) \text{ in } U, \tag{3.8.10}$$

then $t = kt_0$ for some constant $k > 0$.

The display (3.8.10) can be put into a more convenient form by letting $\varphi = t \circ t_0^{(-1)}$ and setting $u = t_0(x)$, $v = t_0(y)$. Since $\varphi \circ t_0 = t$, we obtain

$$\min[\varphi(u) + \varphi(v), \varphi(t_0(0))] = \varphi(u + v), \quad \text{for all } (u, v) \text{ in } t_0^*(U), \tag{3.8.11}$$

where

$$t_0^*(U) = \{(t_0(x), t_0(y)) : (x, y) \text{ in } U\}, \tag{3.8.12}$$

which yields

Lemma 3.8.8 *Let t_0 be a generator of a continuous Archimedean t-norm T_0. Let $\varphi : [0, t_0(0)] \to \mathbb{R}^+$ be a continuous, strictly increasing function with $\varphi(0) = 0$. Let $t = \varphi \circ t_0$ and T be the t-norm generated by t. Let U be a subset of I^2 and $t_0^*(U)$ the set defined in (3.8.12). Then $T = T_0$ on U if and only if (3.8.11) holds.*

Lemma 3.8.9 *The set $U \subseteq I^2$ is a set of uniqueness for $\mathcal{T}_{Ar}$ if and only if, for each continuous and strictly decreasing function $t_0 : I \to \mathbb{R}^+$, the only continuous and strictly increasing solutions φ of (3.8.11) satisfying $\varphi(0) = 0$ are the functions given by*

$$\varphi(u) = ku, \quad \text{for } 0 \le u \le t_0(0), \quad \text{where } k > 0.$$

As a first application of Lemmas 3.8.8 and 3.8.9 we establish a variant of a result given in [Kimberling (1973), Theorem 5]. Specifically, we show that an "arbitrarily thin wedge" along the upper boundary of I^2 is a set of uniqueness for $\mathcal{T}_{Ar}$.

Theorem 3.8.10 *Let $\kappa : (0, 1) \to (0, 1)$ be a continuous, non-decreasing function such that $0 \le \kappa(0+) < 1$ and $\kappa(1-) = 1$. Then the open, connected region*

$$K = \{(x, y) : 0 < x < 1 \text{ and } \kappa(x) < y < 1\}$$

is a set of uniqueness for $\mathcal{T}_{Ar}$.

Proof. Let T and T_0 be in $\mathcal{T}_{Ar}$ with generators t and t_0, respectively, and suppose that $T = T_0$ on K. Let $\varphi = t \circ t_0^{(-1)}$ and let $\psi = t_0 \circ \kappa \circ t_0^{(-1)}$. Note that φ is continuous and strictly increasing on $[0, t_0(0)]$ and that ψ is non-decreasing with $\psi(0+) = 0$ and $\lim_{x\to\infty} \psi(x) = \psi(t_0(\kappa(0+))) > 0$.

Applying Lemma 3.8.8, we find that

$$\varphi(u+v) = \varphi(u) + \varphi(v), \quad \text{for all } (u,v) \text{ in } t_0^*(K),$$

where

$$t_0^*(K) = \{(u,v) : 0 < u < t_0(0),\, 0 < v < \min[\psi(u), t_0(0) - u]\}$$

is an open, connected region in the first quadrant (see Figure 3.8.3).

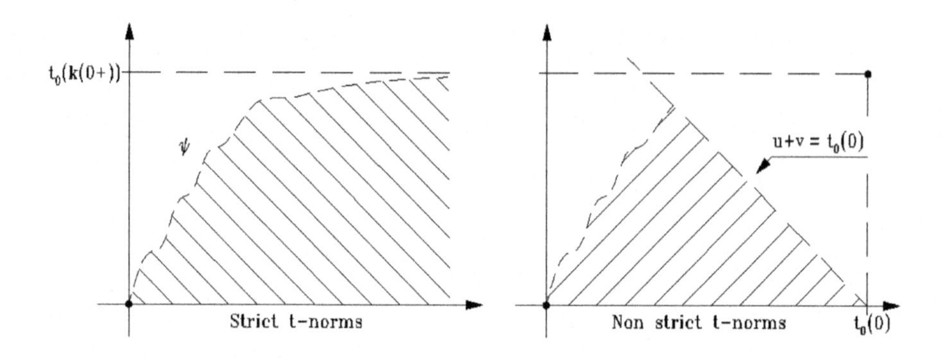

Figure 3.8.3. The region $t_0^*(K)$.

Now, a well-known result regarding Cauchy's equation on restricted domains, first proved in [Daróczy and Losonczi (1967), Satz 4] and later simplified in [Kuzcma (1985), Section VIII-6, Theorem 1] (see also [Kuzcma (1978), Theorem 4.1]) yields that $\varphi(u) = ku$ for some $k > 0$. Thus, by Lemma 3.8.9, $T = T_0$. □

In the remainder of this section we focus our attention on one-dimensional sets that are unions of subsets of the sets D, H_y and $L(z)$ considered in Theorems 3.8.2, 3.8.4 and 3.8.5. We begin by using (3.8.11) and (3.8.12) to reformulate the considerations leading to these theorems as special cases of Lemma 3.8.8.

Lemma 3.8.11 *Let t_0, T_0, φ, t and T be as in Lemma 3.8.8. Then:*

(a) If δ_0 and δ are the diagonals of T_0 and T, respectively, and if, as in (a) of Definition 3.8.1, $d_0 = t_0^{(-1)}\left(\frac{1}{2}t_0(0)\right)$, then $\delta = \delta_0$ on

$J \subseteq [d_0, 1]$ if and only if φ is a solution of the Schröder equation

$$\varphi(2u) = 2\varphi(u), \quad \textit{for all } u \textit{ in } t_0(J) \subseteq \left[0, \frac{1}{2}t_0(0)\right]. \tag{3.8.13}$$

(b) *If, for a fixed y in $(0,1)$, σ_0 and σ are the y-sections of T_0 and T, respectively, if as in (a) of Definition 3.8.3, $c_0 = t_0^{(-1)}[t_0(0) - t_0(y)]$, and if t_0 and t are normalized so that $t_0(y) = t(y) = 1$, then $\sigma = \sigma_0$ on $K \subseteq [c_0, 1]$ if and only if φ is a solution of the Abel equation*

$$\varphi(u+1) = \varphi(u) + 1, \quad \textit{for all } u \textit{ in } t_0(K) \subseteq [0, t_0(0) - 1]\,. \tag{3.8.14}$$

(c) *If, for a fixed z in $(0,1)$, $L_0(z)$ and $L(z)$ are the z-level curves of T_0 and T, respectively, if $L_0(0)$ and $L(0)$ are the boundary curves of $\mathcal{Z}(T_0)$ and $\mathcal{Z}(T)$, and if t_0 and t are normalized to that $t_0(z) = t(z) = 1$, then $L(z) = L_0(z)$ if and only if φ is a solution of the equation*

$$\varphi(1-u) = 1 - \varphi(u), \quad \textit{for } 0 \le u \le 1. \tag{3.8.15}$$

In Lemma 3.8.11, T_0 with generator t_0 and φ are given and T with generator t is constructed. In the other direction, if T_0 and T, with respective generators t_0 and t, are given and if φ is defined on $[0, t_0(0)]$ via $\varphi = t \circ t_0^{(-1)}$, then $\delta = \delta_0$, $\sigma = \sigma_0$, or $L(z) = L_0(z)$ if and only if φ is a solution of (3.8.13), (3.8.14), or (3.8.15), respectively.

Note that when $z = 0$, part (c) of Lemma 3.8.11 is just a restatement of Theorem 2.2.7.

The functional equations (3.8.13) and (3.8.14) are of "iterative type". This means, in particular, that for all n in $\mathbb{Z}$, any solution of (3.8.13) satisfies

$$\varphi(2^n u) = 2^n \varphi(u),$$

provided $2^n u$ belongs to the domain of φ; and any solution of (3.8.14) satisfies

$$\varphi(u+n) = \varphi(u) + n,$$

provided $u + n$ belongs to the domain of φ.

The arguments leading to Theorem 3.8.2, 3.8.4 and 3.8.5 show that each of the equations (3.8.13), (3.8.14) and (3.8.15) has infinitely many non-linear solutions. Thus, as we have already shown, the sets D, H_y and $L(z)$ are not sets of uniqueness for $\mathcal{T}_{Ar}$. However, pairs of these equations often do have a unique common (linear) solution. We begin our presentation of these matters by extending a result due to C. Burgués [1981].

Theorem 3.8.12 *(a) For each z in $(0,1)$, the set $D \cup L(z)$ is a set of uniqueness for $\mathcal{T}_{Ar}$. (b) The set $D \cup L(0)$ is a set of uniqueness for $\mathcal{T}_{Ar} \backslash \mathcal{T}_{St}$.*

Proof. (a) Choose T_0 and T as in Lemma 3.8.11 and assume (without loss of generality) that $t_0(z) = 1$, so that $t_0(0) > 1$. Suppose that $\delta = \delta_0$ and $L(z) = L_0(z)$. Then

$$\begin{cases} \varphi(\frac{1}{2}u) = \frac{1}{2}\varphi(u), & \text{for } 0 \le u \le t_0(0), \\ \varphi(1-u) = 1 - \varphi(u), & \text{for } 0 \le u \le 1. \end{cases} \tag{3.8.16}$$

In view of Lemma 3.8.9 it suffices to show that $\varphi = j_{[0,t_0(0)]}$ is the only continuous and strictly increasing solution of the system (3.8.16). Now (3.8.16) yields $\varphi(0) = 0$, $\varphi(1) = 1 - \varphi(0) = 1$, $\varphi(1/2) = \frac{1}{2}\varphi(1) = 1/2$, $\varphi(1/4) = \frac{1}{2}\varphi(1/2) = 1/4$, $\varphi(3/4) = 1 - \varphi(1/4) = 3/4$. Continuing by induction, assume that $\varphi(m/2^n) = m/2^n$. If $m < 2^n$ then $\varphi(m/2^{n+1}) = \frac{1}{2}\varphi(m/2^n) = m/2^{n+1}$, while if $2^n < m < 2^{n+1}$, then

$$\begin{aligned} \varphi(m/2^{n+1}) &= 1 - \varphi(1 - m/2^{n+1}) = 1 - \varphi[(2^{n+1} - m)/2^{n+1}] \\ &= 1 - \frac{1}{2}\varphi[(2^{n+1} - m)/2^n] = m/2^{n+1} \end{aligned}$$

as well. Thus

$$\varphi(m/2^n) = m/2^n$$

for all dyadic rationals $m/2^n$ in I. Therefore, by continuity, $\varphi(u) = u$ for all u in I. This implies that $t(x) = t_0(x)$ for $z \le x \le 1$, whence $T(x,y) = T_0(x,y)$ whenever $T_0(x,y) \ge z$. Next, for u in $(1, t_0(0)]$, let k be a positive integer such that $u/2^k < 1$. Then by (3.8.16), $\varphi(u) = 2^k\varphi(u/2^k) = u$. Thus $\varphi = j_{[0,t_0(0)]}$.

(b) In this case $t_0(0) = 1$ and the conclusion follows from the first part of the preceding argument. □

In the proof of part (a) of Theorem 3.8.12, the argument showing that $\varphi(u) = u$ for u in $(1, t_0(0)]$ can be modified to show that the removal of any portion of D that lies below $L(z)$ yields a set of non-uniqueness. To this end, let J be a non-trivial subinterval of $[0, z)$, let $[a, b] = t_0(\overline{J})$ and note that $[a, b] \subseteq [1, t_0(0)]$. Now define φ on $[1, t_0(0)]$ as follows: Let $\varphi(u) = u$ for u in $[1, a]$; extend φ to $[a, b]$ so that φ is continuous and increasing, but

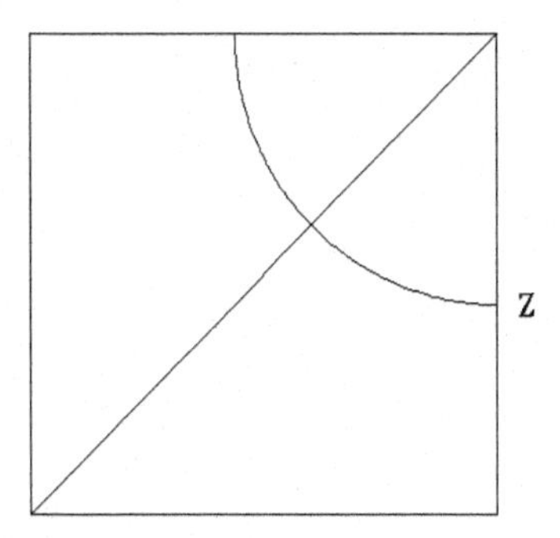

Figure 3.8.4. Set of uniqueness $D \cup L(z)$.

$\varphi \neq j_{[a,b]}$; and extend φ to $[b, t_0(0)]$ via $\varphi(u) = 2\varphi\left(\frac{1}{2}u\right)$. Then φ satisfies the Schröder equation $\varphi\left(\frac{1}{2}u\right) = \frac{1}{2}\varphi(u)$ on $[0, t_0(0)] \backslash [a, b]$, but not on (a, b).

The next two examples show that uniqueness can also be destroyed by removing portions of D that lie above $L(z)$. In the first example, an interval ending at (1,1) is removed; in the second, an interval beginning at $L(z)$.

Example 3.8.13 [Burgués (1981)] Let T_0 be the non-strict t-norm in $\mathcal{T}_{Ar}$ generated by t_0, where $t_0(0) = 1$. Construct a function $\varphi : I \to I$ as follows:

(i) Let φ be strictly increasing from $[1/3, 1/2]$ onto $[1/3, 1/2]$, but $\varphi \neq j_{[1/3,1/2]}$.
(ii) On $[2/3, 1]$ define φ via the Schröder equation $\varphi(u) = 2\varphi(\frac{1}{2}u)$.
(iii) On the remainder of I, define φ via $\varphi(1-u) = 1 - \varphi(u)$.

Now let T be the t-norm generated by $t = \varphi \circ t_0$. It follows at once from Lemma 3.8.11(c) that $\mathcal{Z}(T) = \mathcal{Z}(T_0)$; and it follows from Lemma 3.8.11(a), with $J = \left[t_0^{-1}(1/2), t_0^{-1}(1/3)\right] = [d_0, t_0^{-1}(1/3)]$, that $\delta = \delta_0$ on $[0, t_0^{-1}(1/3)]$. However, since t is not a constant multiple of t_0, clearly $T \neq T_0$.

Thus removing a portion of D above a level set leads to non-uniqueness. (An explicit example with $T_0 = W$ and φ quadratic is presented in Example A.7.1 of Appendix A.) The removed portion may be arbitrarily small: a construction more complicated than the one given above yields a similar example in which $\delta = \delta_0$ on $[0, 1 - \epsilon]$.

Example 3.8.14 Let T_0 be as in Example 3.8.13. To construct a non-strict Archimedean t-norm $T \neq T_0$ for which $\mathcal{Z}(T) = \mathcal{Z}(T_0)$ and $\delta = \delta_0$ on

$[t_0^{-1}(3/8), 1]$, it suffices to construct a strictly increasing function φ from I onto I, $\varphi \neq j_I$, which is a solution of the system

$$\begin{cases} \varphi(2u) = 2\varphi(u), & \text{for } 0 \le u \le 3/8, \\ \varphi(1-u) = 1 - \varphi(u), & \text{for all } u \text{ in } I, \end{cases}$$

and to put $t = \varphi \circ t_0$.

Let $\psi : I \to I$ be any continuous and strictly increasing solution of the Schröder equation

$$\psi(u/4) = \frac{1}{4}\psi(u), \text{ for all } u \text{ in } I,$$

with $\psi(0) = 0$ and $\psi(1) = 1$, but $\psi \neq j_I$. Define φ on [1/3,1/2] via

$$\varphi(u) = \frac{1}{3} + \frac{1}{6}\psi(6u-2), \text{ for } 1/3 \le u \le 1/2,$$

and extend φ to [1/4,1/3] via

$$\varphi(u) = \frac{1}{2}[1 - \varphi(1-2u)] = \frac{1}{3} - \frac{1}{12}\psi(4-12u), \text{ for } 1/4 \le u \le 1/3.$$

Now extend φ from [1/4,1/2] to [0,1/4] via $\varphi(u/2) = \frac{1}{2}\varphi(u)$, recursively, and $\varphi(0) = 0$; and finally from [0,1/2] to I via $\varphi(u) = 1 - \varphi(1-u)$. This φ satisfies the above system since we have $\varphi(2u) = 2\varphi(u)$ when $1/4 \le u \le 3/8$.

(We conjecture that $\varphi = j_I$ is the only solution of the above system when 3/8 is replaced by any larger number, i.e., when a smaller interval starting at $L(z)$ is removed from D.)

The next example shows that a denumerable union of level sets need not be a set of uniqueness for $\mathcal{T}_{Ar}$.

Example 3.8.15 Let T_0 be a strict t-norm and z_1 an element of (0,1). Choose the generator t_0 so that $t_0(z_1) = 1$. Let $\psi : I \to I$ be a continuous and strictly increasing function satisfying $\psi(1-u) = 1-\psi(u)$ and $\psi(0) = 0$, but $\psi \neq j_I$. Define $\varphi : \mathbb{R}^+ \to \mathbb{R}^+$ via

$$\varphi(u) = \lfloor u \rfloor + \psi(< u >),$$

where $\lfloor u \rfloor$ is the greatest integer $\le u$ and $< u >$ is the fractional part of u. Note that for all n in $\mathbb{N}$, $\varphi(u+n) = \varphi(u) + n$ for all u in I, $\varphi(n) = n$, and $\varphi(n-u) = n - \varphi(u)$ for $0 \le u \le n$. Now let $\{z_n\}$ be the decreasing sequence given by $z_n = t_0^{-1}(n)$ for n in $\mathbb{N}$, and let T be the strict t-norm generated by $t = \varphi \circ t_0$. Then $T \neq T_0$, while for each fixed n and all x in $[z_n, 1]$,

$$\begin{aligned} t^{-1}(n - t(x)) &= t_0^{-1} \circ \varphi^{-1}(n - \varphi \circ t_0(x)) \\ &= t_0^{-1} \circ \varphi^{-1} \circ \varphi(n - t_0(x)) = t_0^{-1}(n - t_0(x)), \end{aligned}$$

so that $t(x) + t(y) = n$ if and only if $t_0(x) + t_0(y) = n$, i.e., $L(z_n) = L_0(z_n)$ for all n in $\mathbb{N}$.

When it comes to contours and sections, there is an even stronger negative result.

Theorem 3.8.16 *For each y in $(0, 1)$ and each z in $[0, 1)$, the set $H_y \cup L(z)$ is not a set of uniqueness for T_{Ar}.*

Proof. Fix y and z as above, let T_0 be an Archimedean t-norm with generator t_0 normalized so that $t_0(z) = 1$, and let $a = t_0(y)$. It suffices to construct solutions φ of the system

$$\begin{cases} \varphi(1 - u) = 1 - \varphi(u), \text{ for all } u \text{ in } I, & (3.8.17a) \\ \varphi(u + a) = \varphi(u) + a, \text{ for all } u \text{ in } \mathbb{R}^+, & (3.8.17b) \end{cases}$$

such that $\varphi(0) = 0$, while $\varphi \neq j_{[0, t_0(0)]}$. For then the t-norm T generated by $t = \varphi \circ t_0$ is distinct from T_0, but has $L(z) = L_0(z)$, by (3.8.15), and $\sigma = \sigma_0$ for the given y-section, since – on noting that $t(y) = \varphi(t_0(y)) = \varphi(a) = a = t_0(y)$ – for all x in I such that $\sigma_0(x) > 0$ we have

$$\begin{aligned} \sigma_0(x) &= t_0^{(-1)}[t_0(x) + t_0(y)] = t^{(-1)} \circ t \circ t_0^{(-1)}[t_0(x) + a] \\ &= t^{(-1)}[\varphi(t_0(x) + a)] = t^{(-1)}[\varphi(t_0(x)) + a] \\ &= t^{(-1)}[t(x) + t(y)] = \sigma(x). \end{aligned}$$

As regards the construction of φ, there are two cases.

(1) If $y \leq z$, i.e., if $a \geq 1$ then, on I, take φ to be a continuous increasing solution of (3.8.17a); on $[1, a]$, let $\varphi(u) = u$; and then use (3.8.17b) to extend φ from $[0, a]$ to $\mathbb{R}^+$.

(2) If $y > z$, i.e., if $a < 1$, choose n so that $1/(n + 1) \leq a < 1/n$ and let $b = (1 - na)/2$. On $[a, b]$, let $\varphi(0) = 0$, $\varphi(b) = b$ and let φ be strictly increasing but different from $j_{[0,b]}$; extend φ to $[b, 2b]$ via reflection in the point (b, b), so that $\varphi(u) = 2b - \varphi(2b - u)$ for u in

$[b, 2b]$; let $\varphi(u) = u$ for u in $[2b, a]$; and finally extend φ from $[0, a]$ to $\mathbb{R}^+$ via (3.8.17b). It is easily seen that φ satisfies (3.8.17a). □

There is a similar result concerning pairs of sections which is a consequence of

Lemma 3.8.17 *Let T_0 be a strict t-norm with generator t_0 and let y_1, y_2 be distinct points in $(0, 1)$. Then T_0 is uniquely determined by the y-sections σ_{y_1} and σ_{y_2} if and only if the ratio $t_0(y_2)/t_0(y_1)$ is irrational.*

Proof. Normalize t_0 so that $t_0(y_1) = 1$, let $t_0(y_2) = a$, and let $t = \varphi \circ t_0$, where $\varphi : \mathbb{R}^+ \to \mathbb{R}^+$ is continuous and strictly increasing. Then by Lemmas 3.8.8 and 3.8.11(b), t generates a strict t-norm T having the y-sections σ_{y_1} and σ_{y_2} if and only if, for all u in $\mathbb{R}^+$,

$$\begin{cases} \varphi(u+1) = \varphi(u) + 1, \\ \varphi(u+a) = \varphi(u) + \varphi(a). \end{cases}$$

Suppose first that a is irrational and let m and n be positive integers such that $m/n < a < (m+1)/n$. Then, since $\varphi(0) = 0$,

$$m = \varphi(m) < \varphi(na) = n\varphi(a) < \varphi(m+1) = m+1,$$

i.e., $m/n < \varphi(a) < (m+1)/n$, whence $\varphi(a) = a$. It follows that $\varphi(na) = na$ and that $\varphi(na - m) = na - m$ whenever $na - m > 0$. Thus $\varphi(< na >) = < na >$; and since the set $\{< na >: n \text{ in } N\}$ is dense in I [Niven (1956)] and since φ is continuous, $\varphi = j_I$. Applying (3.8.14) yields $\varphi = j_{\mathbb{R}^+}$ and $T = T_0$.

If, however, a is rational, say $a = m/n$, then we let φ be strictly increasing from $[0, 1/n]$ onto $[0, 1/n]$, but $\varphi \neq j_{[0,1/n]}$, and extend φ recursively to $\mathbb{R}^+$ via $\varphi(u + 1/n) = \varphi(u) + 1/n$. This yields a solution $\varphi \neq j_{\mathbb{R}^+}$. Hence, in this case, T_0 is not uniquely determined by σ_{y_1} and σ_{y_2}. □

Since, for any distinct y_1 and y_2 in (0,1), one can always choose a strict generator t_0 such that $t_0(y_2)/t_0(y_1)$ is rational, we immediately have

Theorem 3.8.18 *For any distinct y_1 and y_2 in $(0, 1)$, the set $H_{y_1} \cup H_{y_2}$ is not a set of uniqueness for $\mathcal{T}_{Ar}$.*

We now turn to a more interesting and intricate question, namely sets of uniqueness determined by restrictions of diagonals and sections, i.e., subsets of $D \cup H_y$. As Lemma 3.8.11 shows, this question leads directly to the consideration of simultaneous solutions of an Abel equation and a

Schröder equation. An extensive study of such **Abel-Schröder systems**, namely systems of the form

$$\begin{cases} \varphi(f(u)) = \varphi(u) + 1, \\ \varphi(g(u)) = 2\varphi(u), \end{cases} \tag{3.8.18}$$

where f and g are given and φ is to be found, was undertaken in [Darsow and Frank (1983)]. Here we present an outline of the salient results as they apply to our situation. We omit most proofs and many details.

For a fixed y in (0,1), let T_0, T be continuous Archimedean t-norms with diagonals δ_0, δ, and y-sections σ_0, σ, respectively. Let d_0, d, c_0, c be as specified in Definitions 3.8.1 and 3.8.3, respectively, and assume that $y \geq d_0$. Let t_0 and t be the generators of T_0 and T, respectively, for which $t_0(y) = t(y) = 1$. Let φ be the strictly increasing function from $[0, t_0(0)]$ onto $[0, t(0)]$ given by $\varphi = t \circ t_0^{-1}$. Then from (a) and (b) of Lemma 3.8.11 we have that, for $J \subseteq [d_0, 1]$ and $K \subseteq [c_0, 1]$,

$$\delta = \delta_0 \text{ on } J \quad \text{ and } \quad \sigma = \sigma_0 \text{ on } K$$

if and only if φ is a solution of the system

$$\begin{cases} \varphi(2u) = 2\varphi(u), & \text{for all } u \text{ in } t_0(J), \\ \varphi(u+1) = \varphi(u) + 1, & \text{for all } u \text{ in } t_0(K). \end{cases} \tag{3.8.19}$$

When it can be shown that φ must be the identity on $[0, t_0(0)]$, then $T = T_0$ and the corresponding subset of $D \cup H_y$ is a set of uniqueness for $\mathcal{T}_{Ar}$. To address this question, we begin with

Lemma 3.8.19 *In the notation of the preceding paragraph, if $\delta = \delta_0$ on $[y, 1]$ and $\sigma = \sigma_0$ on $[y, 1]$, then $T = T_0$ on $[y, 1]^2$.*

Proof. First note that $J = K = [y, 1]$ and $t_0([y, 1]) = I$. Thus, since $T_0(y, y) = t_0^{-1}(2)$ and $t = \varphi \circ t_0$, in view of Lemma 3.8.8 it suffices to show that $\varphi = j_{[0,2]}$. Arguing as in the proof of Theorem 3.8.12, we show by induction on n that $\varphi(m/2^n) = m/2^n$ for each dyadic rational in I. First $\varphi(0) = 0$, $\varphi(1) = 1$ and for $n = 1$ we have $\varphi(1/2) = \frac{1}{2}\varphi(1) = \frac{1}{2}[\varphi(0) + 1] = \frac{1}{2}$. Next, if $m < 2^n$ then $\varphi(m/2^{n+1}) = \frac{1}{2}\varphi(m/2^n) = m/2^{n+1}$, while if $2^n \leq m < 2^{n+1}$, then

$$\begin{aligned} \varphi(m/2^{n+1}) &= \tfrac{1}{2}\varphi(m/2^n) = \tfrac{1}{2}[\varphi(m/2^n - 1) + 1] \\ &= \tfrac{1}{2}[m/2^n - 1 + 1] = m/2^{n+1}. \end{aligned}$$

Thus, by continuity, $\varphi(u) = u$ for u in I; and for u in $[1,\ 2]$, we have $\varphi(u) = \varphi(u-1) + 1 = u$. □

Theorem 3.8.20 *If T and T_0 are continuous Archimedean t-norms and if either*

(a) $\delta = \delta_0$ on I and $\sigma = \sigma_0$ on $[y, 1]$,

or

(b) $\delta = \delta_0$ on $[y, 1]$ and $\sigma = \sigma_0$ on I,

then $T = T_0$.

The proof of Theorem 3.8.20 is an almost immediate consequence of Lemma 3.8.19. In each case, it is easy to show that φ must be the identity on $[0, t_0(0)]$ – in case (a) by extending from I recursively via the Schröder equation to $[0, 2t_0(d_0)] = [0, t_0(0)]$, and in case (b) via the Abel equation to $[0, t_0(c) + 1] = [0, t_0(0)]$. The corresponding sets of uniqueness are depicted in Figure 3.8.5.

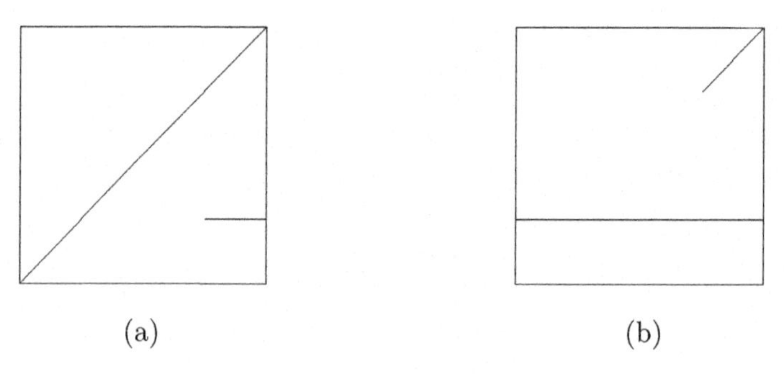

Figure 3.8.5. Sets of uniqueness via Theorem 3.8.20.

Theorem 3.8.21 *If T and T_0 are strict t-norms and if there is an $\epsilon > 0$ for which either*

(a) $\delta = \delta_0$ on I and $\sigma = \sigma_0$ on $[0, \epsilon]$,

or

(b) $\delta = \delta_0$ on $[0, \epsilon]$ and $\sigma = \sigma_0$ on I,

then $T = T_0$.

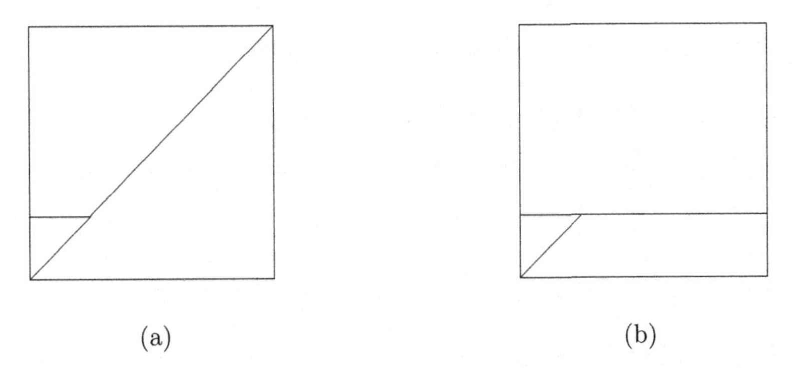

Figure 3.8.6. Sets of uniqueness via Theorem 3.8.21.

The method of proof for the corresponding Abel-Schröder system is not essentially different from the one outlined above. The details of the argument, as well as other examples of sets of uniqueness, are given in [Darsow and Frank (1983)].

Each of the preceding sets of uniqueness is minimal in the sense that the removal of any (interior) interval yields a set on which T is never uniquely determined. In fact, if an interval is removed from either $t_0(J)$ or $t_0(K)$ then the resulting system (3.8.19) always has solutions other than the identity.

The problem of *extension* of restrictions of diagonals and sections goes much deeper. Let J and K be subintervals of I, and suppose that $\delta : J \to I$ and $\sigma : K \to I$ are functions that satisfy all of the requirements of (restrictions of) diagonals and sections. The functions δ and σ are said to be **compatible** if there exists an Archimedean t-norm T whose diagonal and y-section, when restricted to J and K, respectively, are δ and σ, i.e., if δ and σ can be extended to a t-norm, or equivalently, if there exists a generator $t : I \to \mathbb{R}^+$ such that

$$\begin{cases} t(\delta(x)) = 2t(x), & \text{for all } x \text{ in } J, \\ t(\sigma(x)) = t(x) + 1, & \text{for all } x \text{ in } K. \end{cases}$$

It is easy to see that δ and σ are compatible whenever the sets $J \cup \delta(J)$ and $K \cup \sigma(K)$ are non-overlapping intervals. In general, however, conditions for compatibility rest on a detailed study of the construction of solutions of the Abel-Schröder system (3.8.18) and are extremely intricate. (As well as paying careful attention to domains and ranges, one needs to show, for example, that certain sets are dense in certain intervals.) A full discussion, including the solution in several key cases, is presented in [Darsow and Frank (1983)]. A simplified argument for one particular case, but for strict t-norms only, appears in Section 9.3C of [Kuzcma, Choczewski and Ger (1990)]. The general problem is open. An illustration of incompatibility is given in Example A.7.5 of Appendix A.

Abel-Schröder systems, and more generally systems of "conjugacy" equations, i.e., systems of equations of the from $\varphi(f(x)) = g(\varphi(x))$, arise naturally in algebra, analysis and geometry. Our results on the structure of t-norms and the methods employed to obtain them are thus important in a much broader context. Extensive discussions and references to the literature may be found in the survey [Kairies (1997)] and in the book [Kuczma, Choczewski and Ger (1990)].

Let D^* be the second diagonal of I^2, i.e., $D^* = \{(x, 1-x) : x \text{ in } I\}$. For many years it was an open question whether $D \cup D^*$ is a set of uniqueness for strict t-norms. The question was answered in the affirmative in [Bézivin and Tomás (1993)] as a consequence of the following result:

Lemma 3.8.22 *Let β be a strictly decreasing function from I onto I. Let T and T_0 be strict t-norms and, for each integer $k \geq 2$ and each x in I, let x_T^k and $x_{T_0}^k$ denote the k-th T-power and the k-th T_0-power of x, respectively (see (1.3.9)). Suppose that for some $k \geq 2$ and all x in I,*

$$\begin{cases} x_T^k = x_{T_0}^k, \\ T(x, \beta(x)) = T_0(x, \beta(x)). \end{cases} \tag{3.8.20}$$

Then $T = T_0$.

The first step in the proof of Lemma 3.8.22 is to transform the system (3.8.20) to a system of functional equations. Indeed, setting $\varphi = t \circ t_0^{-1}$ and $\psi = t_0 \circ \beta \circ t_0^{-1}$ yields that for all u in $\mathbb{R}^+$

$$\begin{cases} \varphi(ku) = k\varphi(u), \\ \varphi(u + \psi(u)) = \varphi(u) + \varphi(\psi(u)), \end{cases} \tag{3.8.21}$$

where φ and ψ are continuous functions from $\mathbb{R}^+$ onto $\mathbb{R}^+$, with φ strictly increasing and ψ strictly decreasing. A clever argument then yields that the function $\varphi(x)/x$ must be constant. Thus, independent of ψ, the only solutions of the system (3.8.21) are $\varphi = cj_{\mathbb{R}^+}$, and hence $T = T_0$.

In general, since $x_T^k = x_{T_0}^k$ for all x in I and some $k > 2$ neither implies nor is implied by $x_T^2 = x_{T_0}^2$ for all x in I, Lemma 3.8.22 is not a "sets of uniqueness" result. (Indeed, $T = T_0$ on the curve $y = T_0(T_0(x,x), x)$ need not imply that $T(T(x,x),x) = T_0(T_0(x,x),x)$ for all x in I.) However, the special case $k = 2$ yields

Theorem 3.8.23 *Let β be as in Lemma 3.8.22 and let $D_\beta = \{(x, \beta(x)) : x \text{ in } I\}$. If T and T_0 are strict t-norms such that for all x in I,*

$$\begin{cases} T(x,x) = T_0(x,x), \\ T(x,\beta(x)) = T_0(x,\beta(x)), \end{cases} \tag{3.8.22}$$

then $T = T_0$.

Thus $D \cup D_\beta$, and in particular $D \cup D'$, is a set of uniqueness for strict t-norms (see Figure 3.8.7).

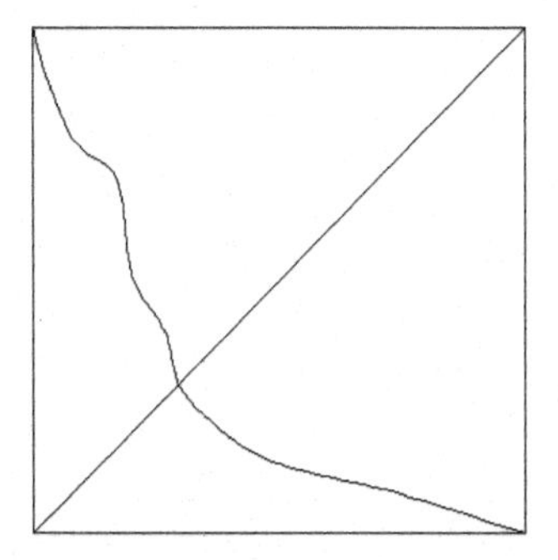

Figure 3.8.7. Sets of uniqueness via Theorem 3.8.23.

Similar methods yield the sets of uniqueness depicted in Figure 3.8.8. These sets are not minimal; and the results are false for non-strict t-norms (see [Bézivin and Tomás (1993)] for details).

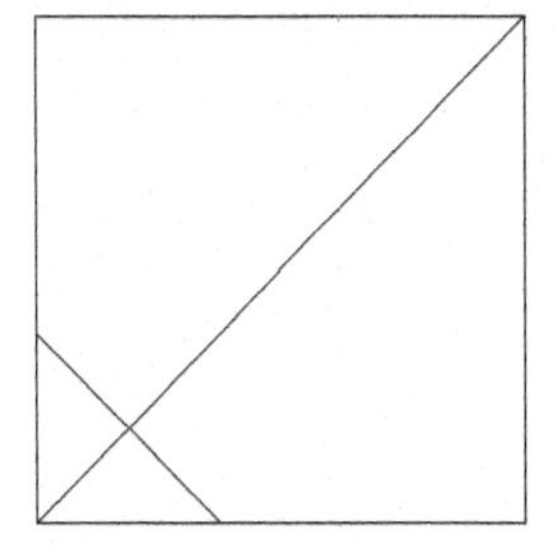

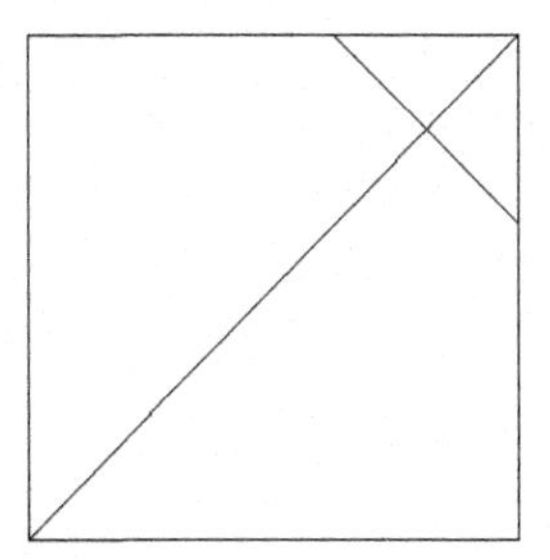

Figure 3.8.8. Sets of uniqueness for strict t-norms.

Chapter 4

Inequalities involving t-norms

4.1 Notions of concavity and convexity

In this section we investigate several notions of concavity and convexity for two-place functions and their connection with associativity. We focus our attention on Schur-concavity and Schur-convexity, which play a fundamental role in the theory of majorization and related topics (see [Marshall and Olkin (1979)]), on quasi-concavity and quasi-convexity, which are important in optimization theory (see, e.g., [de Finetti (1949); Fenchel (1953); Roberts and Varberg (1973); Behringer (1980)]), and on the classical notions of concavity and convexity (see [Hardy, Littlewood and Pólya (1952)]).

The following idea is central to a significant portion of the theory of inequalities [Marshall and Olkin (1979)]:

Definition 4.1.1 Let (z, w) and (x, y) be points in I^2. Then (z, w) is **majorized** by (x, y) if

$$x + y = z + w \quad \text{and} \quad \text{Max}(z, w) \leq \text{Max}(x, y),$$

or, equivalently, if there exists a λ in I such that

$$z = \lambda x + (1 - \lambda)y \quad \text{and} \quad w = (1 - \lambda)x + \lambda y.$$

Observe that (z, w) is majorized by (x, y) if and only if the points (z, w) and (w, z) both lie on the segment joining the points (x, y) and (y, x).

We first consider various notions of concavity.

Definition 4.1.2 A function $T : I^2 \to I$ is **Schur-concave** if $T(z, w) \geq T(x, y)$ whenever (z, w) is majorized by (x, y) or, equivalently, if T satisfies the inequality

$$T(x, y) \leq T(\lambda x + (1 - \lambda)y, (1 - \lambda)x + \lambda y), \tag{4.1.1}$$

for all x, y and λ in I.

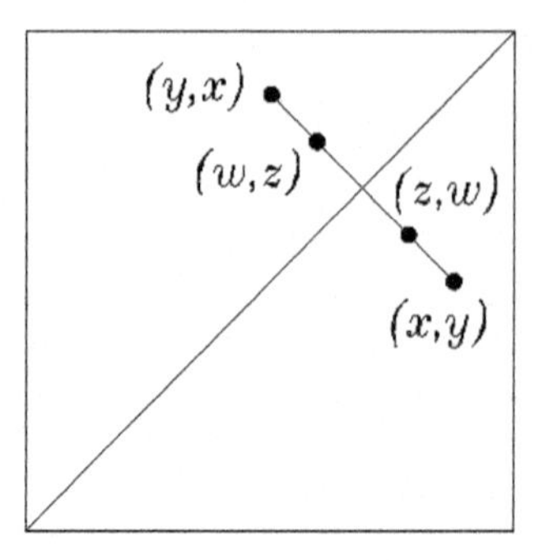

Figure 4.1.1. The point (z,w) is majorized by (x,y).

Since (x,y) is majorized by (y,x) and vice versa, it is immediate that if $T : I^2 \to I$ is Schur-concave then T is commutative. Furthermore, since $((x+y)/2,(x+y)/2))$ is majorized by (x,y), it follows from (4.1.1) that $T : I^2 \to I$ is Schur-concave if and only if, for each a in (0,1), the curve of intersection of the surface $z = T(x,y)$ and the vertical plane $x+y=2a$ descends symmetrically in both directions from its peak at $(a,a,T(a,a))$; or equivalently, if the function $\phi_a : [\max(2a-1,0),\min(2a,1)] \to I$ defined by $\phi_a(x) = T(x,2a-x)$ is non-decreasing when $x < a$ and satisfies $\phi_a(2a-x) = \phi_a(x)$. Consequently, if $T : I^2 \to I$ is Schur-concave, then

$$\sup_{x+y=2a} T(x,y) = \phi_a(a) = \delta_T(a), \tag{4.1.2}$$

for each a in (0,1).

Definition 4.1.3 A function $T : I^2 \to I$ is **quasi-concave** if it satisfies the inequality

$$\mathrm{Min}(T(x,y),T(z,w)) \le T(\lambda x + (1-\lambda)z, \lambda y + (1-\lambda)w), \tag{4.1.3}$$

for all x, y, z, w and λ in I.

In addition, we recall that $T : I^2 \to I$ is **concave** in the usual sense if

$$\lambda T(x,y) + (1-\lambda)T(z,w) \le T(\lambda x + (1-\lambda)z, \lambda y + (1-\lambda)w), \tag{4.1.4}$$

for all x, y, z, w and λ in I; and that if (4.1.4) is assumed to hold only for $\lambda = 1/2$, then T is **midpoint-concave**; and that any continuous midpoint-concave function is concave.

Clearly, if T is concave then T is quasi-concave. Furthermore, on letting $x = w$ and $y = z$ in (4.1.3), it follows at once that if T is commutative and quasi-concave, then T is Schur-concave. However, a Schur-concave function need not be quasi-concave (see Example 4.1.8); and, as is well known, a quasi-concave function, e.g. W, need not be concave.

Quasi-concave and quasi-convex two-place functions were first introduced by B. de Finetti [1949] because, as he showed, a function is quasi-convex, or quasi-concave, if and only if certain unions of its level sets are convex. For our purposes we need a slight variation of this classic result.

Theorem 4.1.4 *Suppose that $T : I^2 \to I$ is non-decreasing in each place and continuous. For $0 \leq z < 1$, let L_z be the function whose graph is the upper boundary curve of the z-level set of T, i.e., let*

$$L_z(x) = \sup\{y \text{ in } I : T(x, y) = z\}, \text{ for all } x \text{ in } I.$$

Then T is quasi-concave if and only if the function L_z is convex for all z in $[0, 1)$.

Proof. Suppose that L_z is convex for each z in [0,1), so that the regions $R(z) = \{(x, y) \text{ in } I^2 \mid T(x, y) \geq z\}$ are convex sets. Choose points $P = (x_1, y_1)$ and $Q = (x_2, y_2)$ in I^2 and let $a = \text{Min}[T(x_1, y_1), T(x_2, y_2)]$. Since P and Q belong to $R(a)$, the entire segment joining P and Q lies in $R(a)$, whence T is quasi-concave.

For the converse, suppose that L_a is not convex for some a, so that the region $R'(a) = \{(x, y) \text{ in } I^2 : T(x, y) > a\}$ is not a convex set. Then there exist P and Q in $R'(a)$ and a point (x, y) on the segment joining P and Q such that $T(x, y) \leq a$, whence T is not quasi-concave. □

Turning to t-norms and copulas, we begin with:

Lemma 4.1.5 *The t-norms Min and W are Schur-concave. If T is a Schur-concave t-norm, then $W \leq T \leq$ Min.*

Proof. The verification of the first statement is immediate. The second follows from the fact that if $(x + y - 1) > 0$, then (x, y) is majorized by $(x + y - 1, 1)$. □

Theorem 4.1.6 *Let T be a strict t-norm with additive generator t. Then the following are equivalent:*

(i) T is Schur-concave;
(ii) T is a copula, i.e., t is convex;
(iii) T is quasi-concave.

Proof. Suppose that T is Schur-concave. Then since (x, y) majorizes $(\frac{x+y}{2}, \frac{x+y}{2})$, for each (x, y) in I^2, we have

$$t^{-1}(t(x)+t(y)) = T(x,y) \leq T\left(\frac{x+y}{2}, \frac{x+y}{2}\right) = t^{-1}\left(2t\left(\frac{x+y}{2}\right)\right).$$

But t^{-1} is decreasing, so that

$$t(x)+t(y) \geq 2t\left(\frac{x+y}{2}\right),$$

whence t, being continuous and midpoint convex, is convex and (i) implies (ii).

Next, if T is a copula, then t is convex so that

$$\begin{aligned}
&t(\lambda x+(1-\lambda)y)+t(\lambda z+(1-\lambda)w) \\
&\leq \lambda t(x)+(1-\lambda)t(y)+\lambda t(z)+(1-\lambda)t(w) \\
&= \lambda(t(x)+t(z))+(1-\lambda)(t(y)+t(w)) \\
&\leq \text{Max}(t(x)+t(z), t(y)+t(w)).
\end{aligned}$$

Applying t^{-1} yields (4.1.3), i.e., that T is quasi-concave. Thus (ii) implies (iii).

Finally, since T is commutative, (iii) implies (i). □

Theorem 4.1.7 *Every associative copula C is quasi-concave.*

Proof. If C is an ordinal sum, then the graph of each L_z consists of a vertical segment, a boundary curve of an Archimedean summand, and a horizontal segment (or just the linear segments or just the boundary curve). Thus, in view of Theorem 4.1.4, we may assume that C is Archimedean.

Let t be a generator of C. For any level z in $[0,1)$, the function L_z is determined by the equation $L_z(x) = t^{-1}(t(z) - t(x))$. Since t is convex, we have

$$t(z) - t\left(\frac{x+y}{2}\right) \geq t(z) - \frac{t(x)+t(y)}{2}$$

$$= \frac{[t(z) - t(x)] + [t(z) - t(y)]}{2}.$$

Therefore, since t^{-1} is decreasing and convex, it follows that

$$\begin{aligned} L_z\left(\frac{x+y}{2}\right) &= t^{-1}\left[t(z) - t\left(\frac{x+y}{2}\right)\right] \\ &\leq t^{-1}\left(\frac{[t(z) - t(x)] + [t(z) - t(y)]}{2}\right) \\ &\leq \frac{1}{2}\left[t^{-1}\left[t(z) - t(x)\right] + t^{-1}[t(z) - t(y)]\right] \\ &= \frac{L_z(x) + L_z(y)}{2}. \end{aligned}$$

Thus L_z is convex and the conclusion follows from Theorem 4.1.4. □

Since an associative copula is commutative, it follows at once that every associative copula is Schur-concave. But associativity in Theorem 4.1.7 is not necessary: for example, the non-associative copula $(P + \mathrm{Min})/2$ has convex level curves.

Every quasi-concave t-norm is clearly Schur-concave, i.e., for non-strict t-norms, (iii) of Theorem 4.1.6 still implies (i). The converse is false, as the following example shows.

Example 4.1.8 Let T be the non-strict Archimedean t-norm generated by

$$t(x) = \begin{cases} \frac{4}{3}\left(2 - \frac{1}{1-x}\right), & 0 \leq x \leq 1/3, \\ 1 - x, & 1/3 \leq x \leq 1. \end{cases}$$

It follows easily that the function L_0 is concave on $[0, 1/3]$ (and also on $[1/3, 1]$) whence T is not quasi-concave; but T is Schur-concave, since a straightforward but somewhat lengthy computation yields that, for each a in $(0, 1)$, the condition described in the paragraph following Definition 4.1.2 is satisfied.

The next example shows that a non-strict quasi-concave t-norm need not be a copula, i.e., that in general neither (iii) nor (i) of Theorem 4.1.6 implies (ii).

Example 4.1.9 Let T be the t-norm generated by $t(x) = 1 + \cos \pi x$. Since t is not convex, T is not a copula: but T is quasi-concave, since for each a in $(0,1)$,

$$L_a(x) = \frac{1}{\pi} \arccos(\cos \pi a - \cos \pi x - 1),$$

and a calculation shows that $L_a''(x) \geq 0$ whence each level curve L_a is convex.

In addition, for the sake of completeness, we note that a non-commutative copula cannot be Schur-concave and that the non-associative copula $(W + \text{Min})/2$ has non-convex level curves and thus is not quasi-concave.

While there is an abundance of Schur-concave and quasi-concave t-norms, there is only one concave t-norm [Agell (1984)].

Theorem 4.1.10 *Let $T : I^2 \to I$ satisfy (1.3.1b) and (1.3.2). Then T is midpoint-concave if and only if $T = Min$.*

Proof. It is easily verified that $T = \text{Min}$ satisfies (4.1.4). Conversely, from (4.1.4) with $\lambda = \frac{1}{2}$ and the fact that $T \leq \text{Min}$, it follows by induction that, for each positive integer n,

$$\delta_T\left(\sum_{i=1}^{n} x_i/2^i\right) = \sum_{i=1}^{n} x_i/2^i,$$

for all $x_1, \ldots, x_n$ in {0,1}. Hence $T(x,x) = x$ for all dyadic rationals x in I, and the result follows from the monotonicity of T and Lemma 1.3.6. □

Closely related to concavity is the notion of superadditivity: a t-norm T is **superadditive** if

$$T(x,y) + T(z,w) \leq T(x+z, y+w),$$

whenever x, y, z, w in I are such that $\text{Max}(x+z, y+w) \leq 1$. It is an open question whether every strict superadditive t-norm is a copula. This is false for non-strict Archimedean t-norms; see Example A.8.8 of Appendix A.

We now consider the analogous notions of convexity.

Definition 4.1.11 Let $T : I^2 \to I$. Then

(a) T is **Schur-convex** if $T(z,w) \leq T(x,y)$ whenever (z,w) is majorized by (x,y) or, equivalently, if

$$T(\lambda x + (1-\lambda)y, (1-\lambda)x + \lambda y) \leq T(x,y), \tag{4.1.5}$$

for all x, y and λ in I.

(b) T is **quasi-convex** if

$$T(\lambda x + (1-\lambda)z, \lambda y + (1-\lambda)w) \leq \mathrm{Max}(T(x,y), T(z,w)), \tag{4.1.6}$$

for all x, y, z, w and λ in I.

(c) T is **convex** if

$$T(\lambda x + (1-\lambda)z, \lambda y + (1-\lambda)w) \leq \lambda T(x,y) + (1-\lambda)T(z,w), \tag{4.1.7}$$

for all x, y, z, w and λ in I.

(d) T is **midpoint convex** if the inequality (4.1.7) holds for all x, y, z, w in I and $\lambda = 1/2$.

As before, it follows easily that if T is Schur-convex, then T is commutative; that T is Schur-convex if and only if, for each a in (0,1), the curve of intersection of the surface $z = T(x, y)$ and the vertical plane $x + y = 2a$ ascends symmetrically in both directions from its trough at $(a, a, T(a, a))$ (i.e., if the function ϕ_a defined earlier is non-increasing when $x < a$, etc.), whence

$$\inf_{x+y=2a} T(x,y) = \delta_T(a);$$

that if T is quasi-convex and commutative, then T is Schur-convex; and that if T is continuous and non-decreasing, then T is quasi-convex if and only if, for each z in [0,1), the upper boundary curve of the z-level set of T is concave.

Turning to t-norms, and noting as before that here convexity implies quasi-convexity implies Schur-convexity, we begin with:

Lemma 4.1.12 *The t-norms Z and W are Schur-convex. If T is a Schur-convex t-norm, then $Z \leq T \leq W$; and if, in addition, T is in $\mathcal{T}_{Co}$, then T is in $\mathcal{T}_{Ar} \backslash \mathcal{T}_{St}$.*

Proof. The first two statements follow as in Lemma 4.1.5, and the third from the fact that if $T \leq W$, then T has no interior idempotents and thus cannot be an ordinal sum. □

Theorem 4.1.13 *There are no Schur-convex, quasi-convex or convex strict t-norms. W is the only Schur-convex, quasi-convex or convex copula.*

Lemma 4.1.14 *Let T be in $\mathcal{T}_{Ar} \backslash \mathcal{T}_{St}$. Then T is Schur-convex if and only if any generator t of T satisfies the inequality*

$$t(x) + t(y) \leq t(\lambda x + (1 - \lambda)y) + t((1 - \lambda)x + \lambda y), \tag{4.1.8}$$

for all x, y and λ in I for which the right-hand side is less than $t(0)$.

Theorem 4.1.15 *If T is in $\mathcal{T}_{Ar} \backslash \mathcal{T}_{St}$ and has a concave generator, then T is Schur-convex.*

The converse is false as the following example shows.

Example 4.1.16 The t-norm $T_{1/2}$ of Family (2.6.13) given by

$$T(x, y) = \begin{cases} \left[\sqrt{xy} - \sqrt{(1-x)(1-y)}\,\right]^2, & x + y \geq 1, \\ 0, & x + y \leq 1, \end{cases}$$

and generated by $t(x) = \arccos \sqrt{x}$, is Schur-convex, but t is not concave.

The t-norm of Example 4.1.16 is quasi-convex since, as a straightforward calculation shows, its level curves are concave. Example A.8.4 of Appendix A is Schur-convex but not quasi-convex. A simple example of an Archimedean t-norm which is quasi-convex but not convex is the member of Family (2.6.1) with $\alpha = -2$, namely,

$$T(x, y) = \left[\text{Max}(x^2 + y^2 - 1, 0)\right]^{1/2}.$$

To illustrate the preceding discussion further, consider the conical t-norms studied in Section 3.6. These are all non-strict and Archimedean. As shown in Theorem 3.6.4, they form a one-parameter family $\{T_c : c > 0\}$, where T_c is generated by $t_c(x) = (1 - x)^{1/c}$ and where the boundary curve of $\mathcal{Z}(T_c)$ is given by $g_c(x) = 1 - \left[1 - (1 - x)^{1/c}\right]^c$. It is immediate that $t_c'' \geq 0$ for $c \geq 1$ and $t_c'' < 0$ for $c < 1$. Moreover, the calculation carried out in connection with (3.6.2) shows that the same is true of g_c, so that the level curves of T_c are convex when $c \geq 1$ and concave when $c \leq 1$. It therefore follows that T_c is quasi-concave (and therefore a copula) when $c \geq 1$, and quasi-convex when $c \leq 1$.

It is an interesting open problem to characterize continuous convex t-norms by way of their additive generators. Here we have the following negative result which is motivated by the inequality $Z \leq T \leq W$ and the observation that W is not continuously differentiable.

Definition 4.1.17 A continuous Archimedean t-norm T with generator t is **smooth** if t' and t'' exist on (0,1), $t'(x) \neq 0$ for all x in $(0,1)$, $\lim_{x\searrow 0} t'(x) = -\infty$, and t'' is bounded in some left-neighborhood $(1-\epsilon, 1)$ of 1.

Theorem 4.1.18 *Smooth convex t-norms do not exist.*

Proof. We outline the key steps of the proof given in [Alsina and Tomás (1988)]. Let T be a smooth convex t-norm. Then, since $T \leq W$, T must be non-strict, and its generator t for which $t(0) = 1$ is concave on $\left[t^{-1}(1/2), 1\right]$. By virtue of the extension of Theorem 2.2.12 to non-strict t-norms, as described at the end of Section 2.2, T has first and second partial derivatives. A well-known result on convex two-place functions (see [Hardy, Littlewood, and Pólya (1952)]) yields that

$$D_{11}T(x, y) \geq 0, \tag{4.1.9}$$

and

$$D_{11}T(x, y) D_{22}T(x, y) - [D_{12}T(x, y)]^2 \geq 0, \tag{4.1.10}$$

for all (x, y) in $(0, 1)^2$.

Define $g : (0, 1) \to \mathbb{R}$ by $g(u) = t''(t^{-1}(u))/[t'(t^{-1}(u))]^2$. Then (4.1.9), (4.1.10) and (2.2.19) can be combined to give

$$g(u) \leq g(u + v), \tag{4.1.11}$$

and

$$g(u)g(v) \geq (g(u) + g(v))\, g(u + v), \tag{4.1.12}$$

for all u, v in (0,1) with $u + v < 1$. Moreover, the concavity of t yields

$$g(u) \leq 0, \tag{4.1.13}$$

for all u in $(0, \frac{1}{2}]$. Since t'' is bounded on a left-neighborhood of 1 and, as it is easily seen, $\lim_{x\nearrow 1} t'(x) = -\infty$, we have

$$\lim_{u\searrow 0} g(u) = \lim_{x\nearrow 1} \frac{t''(x)}{t'(x)^2} = 0. \tag{4.1.14}$$

Combining (4.1.11), (4.1.12), (4.1.13) and (4.1.14) yields that g must be identically zero, i.e., $t''(t^{-1}(u)) = 0$ for all u in (0,1), whence $t(x) = ax + b$, in contradiction to the condition $\lim_{x\searrow 0} t'(x) = -\infty$. $\square$

Finally, we note that for s-norms, the above concavity (resp., convexity) relations are equivalent to the corresponding convexity (resp., concavity) relations for the associated t-norms, and that the results translate accordingly.

4.2 The dominance relation

In this section we consider a binary relation on two-place functions which first arose in the study of products of probabilistic metric spaces [Tardiff (1980), (1984)], occurs in the same context in the study of ordinary metric spaces, and which has since turned out to be more prevalent and intriguing than originally anticipated.

Definition 4.2.1 Let $(S, \leq)$ be a partially ordered set, and let A and B be binary operations on S. Then A **dominates** B, and we write $A >> B$, if for all x_1, x_2, y_1, y_2 in S,

$$A\left(B(x_1, y_1), B(x_2, y_2)\right) \geq B\left(A(x_1, x_2), A(y_1, y_2)\right). \tag{4.2.1}$$

Many classical inequalities are special cases of (4.2.1). For example, if $S = \mathbb{R}$ and A is addition, then (4.2.1) becomes

$$B(x_1, y_1) + B(x_2, y_2) \geq B(x_1 + x_2, y_1 + y_2), \tag{4.2.2}$$

which says that B is subadditive. Similarly, if B is addition, then (4.2.1) says that A is superadditive. If $A(x, y) = \frac{1}{2}(x + y)$, then (4.2.1) says that B is midpoint convex; and, with $B(x, y) = \frac{1}{2}(x + y)$, that A is midpoint concave. Lastly, if $S = \mathbb{R}^+$, A is addition and $B(x, y) = (x^p + y^p)^{1/p}$, where $p \geq 1$, then (4.2.1) is the classical Minkowski inequality.

When equality holds, (4.2.1) is a generalized bisymmetry equation; and if, moreover, $A = B$, then (4.2.1) is the classical bisymmetry equation (2.8.5). For details see [Aczél (1966)].

To illustrate further, suppose (S_1, d_1) and (S_2, d_2) are metric spaces and that one wishes to construct a metric d on their Cartesian product $S_1 \times S_2$. It is natural to assume that, if p_1, q_1 are points in S_1 and p_2, q_2 points in S_2, then the distance between (p_1, p_2) and (q_1, q_2) is given by

$$d\left((p_1, p_2), (q_1, q_2)\right) = K\left(d_1(p_1, q_1), d_2(p_2, q_2)\right), \tag{4.2.3}$$

where K is a suitable binary operation on $\mathbb{R}^+$. It follows that d will be a semimetric which agrees with d_1 on S_1 and d_2 on S_2 whenever K has 0 as identity and $K(x, y) > 0$ when $x > 0$ or $y > 0$; that d will be isometric to the corresponding metric induced on $S_2 \times S_1$ whenever K is commutative; and – most important – that d will satisfy the triangle inequality whenever K is dominated by addition (i.e., subadditive) and non-decreasing. If, furthermore, K is associative, then (4.2.3) can be used to define the product of three or more metric spaces in a consistent fashion. Finally, to avoid pathologies, it is convenient to assume that K is continuous. In short, K must belong to $\mathcal{G}_{Co}(\mathbb{R}^+)$. (See [Motzkin (1936); Bohnenblust (1940); Tardiff (1980); Schweizer and Sklar (1983), (2005)].)

Theorem 4.2.2 *Let $(S, \leq)$ be a partially ordered set and let A and B be binary operations on S, having a common identity element e.*

(a) If $A >> B$, then $A \geq B$, i.e., if A dominates B, then, in the usual pointwise ordering, A is stronger than B.
(b) The dominance relation is antisymmetric.
(c) The dominance relation is reflexive, i.e., $A >> A$, if and only if A is associative and commutative.

Proof. Suppose $A >> B$. Setting $y_1 = x_2 = e$ in (4.2.1) establishes (a) and then (b) follows as an immediate consequence. To prove (c), suppose $A >> A$. Then, writing xy for $A(x, y)$ and using (4.2.1) twice, we have

$$(x_1y_1)(x_2y_2) \geq (x_1x_2)(y_1y_2) \geq (x_1y_1)(x_2y_2),$$

whence $(x_1y_1)(x_2y_2) = (x_1x_2)(y_1y_2)$. Now setting $x_2 = e$ shows that A is associative. Commutativity follows from

$$\begin{aligned} x_1x_2 &= (ex_1)(x_2e) \geq (ex_2)(x_1e) = x_2x_1 \\ &= (ex_2)(x_1e) \geq (ex_1)(x_2e) = x_1x_2. \end{aligned}$$

Conversely, if A is commutative and associative, then

$$\begin{aligned} (x_1y_1)(x_2y_2) &= ((x_1y_1)x_2)\, y_2 = (x_1(y_1x_2))\, y_2 \\ &= (x_1(x_2y_1))\, y_2 = ((x_1x_2)y_1)\, y_2 \\ &= (x_1x_2)(y_1y_2). \end{aligned}$$

This establishes (c) and completes the proof. □

The hypotheses of Theorem 4.2.2 do not imply that the dominance relation is transitive, i.e., that the relation is a partial order, as the following example, due to H. Sherwood, shows:

Example 4.2.3 Let $S = \{0, 1, 2\}$ be linearly ordered by $0 < 1 < 2$ and let F, G, H be the binary operations on S defined by the multiplication tables:

F	0	1	2
0	0	1	2
1	1	0	2
2	2	2	2

G	0	1	2
0	0	1	2
1	1	1	2
2	2	2	2

H	0	1	2
0	0	1	2
1	1	2	2
2	2	2	2

.

Then F, G, H are commutative and associative operations on S, having 0 as a common identity and such that H dominates G, G dominates F, but H does not dominate F, since $F(H(1,1), H(1,0)) = F(2,1) = 2$ while $H(F(1,1), F(1,0)) = H(0,1) = 1$.

Turning to t-norms, we begin with

Lemma 4.2.4 *For each t-norm T, $\mathrm{Min} >> T >> Z$.*

Proof. For all a, b, c, d in I, we have $T(a,b) \geq T(\mathrm{Min}(a,c), \mathrm{Min}(b,d))$ and $T(c,d) \geq T(\mathrm{Min}(a,c), \mathrm{Min}(b,d))$ whence

$$\mathrm{Min}(T(a,b), T(c,d)) \geq T(\mathrm{Min}(a,c), \mathrm{Min}(b,d)),$$

i.e., $\mathrm{Min} >> T$; similarly, $T >> Z$. □

The next result, which is basic to the subsequent discussion, is due to R. Tardiff [1980].

Theorem 4.2.5 *Let T_1 and T_2 be strict t-norms with additive generators t_1 and t_2, respectively, and let $h = t_1 \circ t_2^{-1}$. Then $T_1 >> T_2$ if and only if*

$$h^{-1}\left(h(u_1 + v_1) + h(u_2 + v_2)\right) \leq h^{-1}\left(h(u_1) + h(u_2)\right) + h^{-1}(h(v_1) + h(v_2)). \quad (4.2.4)$$

Furthermore, in this case h is convex on $\mathbb{R}^+$; and if H is the function defined on $\mathbb{R}^+ \times \mathbb{R}^+$ by

$$H(u,v) = h^{-1}(h(u) + h(v)), \quad (4.2.5)$$

then H is subadditive or, in other words, addition dominates H.

Proof. The inequality (4.2.4) is a straightforward verification. By Theorem 4.2.2, $T_1 >> T_2$ implies $T_1 \geq T_2$ whence, by Theorem 2.2.2, we have $h^{-1} = t_2 \circ t_1^{-1}$ is subadditive and, since $h^{-1}(0) = 0$, h^{-1} is concave. Thus h, being increasing, is convex. The verification of the subadditivity of H is again straightforward.

Note: The convexity of h can also be established directly from (4.2.4), i.e., for functions not necessarily of the form $t_1 \circ t_2^{-1}$: simply assume, without loss of generality, that $0 \leq x \leq y$ and let $u_1 = h^{-1}(x)$, $u_2 = h^{-1}(\frac{y-x}{2})$, $v_1 = h^{-1}(\frac{x+y}{2}) - h^{-1}(x)$ and $v_2 = 0$. □

Using Theorem 4.2.5 it is easy to show that $T_1 \geq T_2$ does not imply $T_1 >> T_2$.

Example 4.2.6 Let $t_1(x) = x^{-1} - 1$ and let

$$t_2(x) = \begin{cases} (2x)^{-1}, & 0 < x \leq \frac{1}{2}, \\ t_1(x), & \frac{1}{2} \leq x \leq 1. \end{cases}$$

Then, $t_2 \circ t_1^{-1}$, being concave and equal to 0 at 0, is subadditive, so that $T_1 \geq T_2$. But since (4.2.4) fails when $u_1 = u_2 = v_2 = 1$ and $v_1 = 1/2$, T_1 does not dominate T_2.

The inequality (4.2.4), which may be viewed as a generalized Minkowski inequality (let $h(u) = u^p$ with $p \geq 1$) was studied in depth in [Mulholland (1950)]. Indeed, it is now commonly referred to as "Mulholland's inequality". In his paper, Mulholland (assuming differentiability) obtained a necessary condition that h be convex. As regards the more difficult question of sufficiency, he showed that the following condition implies (4.2.4).

(i) The function $h : \mathbb{R}^+ \to \mathbb{R}^+$ is continuous, strictly increasing and convex, with $h(0) = 0$, and $\log h$ is a convex function of $\log x$, i.e., the function $\psi : \mathbb{R} \to \mathbb{R}$ given by $\psi = \log \circ h \circ \exp$ is convex.

In addition, he showed that (i) is equivalent to the condition

(ii) There exists a continuous, strictly increasing, convex function $\psi : \mathbb{R} \to \mathbb{R}$ such that $h(u) = u \exp[\psi(\log u)]$.

(A somewhat simpler proof of the sufficiency of Condition (i) and of the equivalence of (i) and (ii) is given in [Kuczma (1985)].) Mulholland also showed that if

$$h(u) = \int_0^\infty u^{v+1} dF(v),$$

where F is any increasing function such that the integral converges for all $u > 0$, then h satisfies Condition (i). A number of years later, Tardiff [1980, 1984] showed that (4.2.4) also follows from the condition

(iii) The function $h : \mathbb{R}^+ \to \mathbb{R}^+$ is differentiable, strictly increasing and convex, with $h(0) = 0$, and the function $\eta : \mathbb{R} \to \mathbb{R}$ which is given by $\eta = \log \circ h' \circ \exp$ is convex.

More recently, W. Jarczyk and J. Matkowski [2002] have shown that Mulholland's condition (i) implies Tardiff's condition (iii), but that the converse is false.

Example 4.2.7 Let $f : (0, \infty) \to (0, \infty)$ be increasing, differentiable, and such that $f(u)/u$ is integrable; and let $h : (0, \infty) \to (0, \infty)$ be defined by

$$h(x) = x \exp \int_0^x [f(u)/u] du.$$

Then h satisfies (i). But for a suitable choice of f, e.g.,

$$f(u) = \frac{u + \sin u}{u + \sin u + 1},$$

h fails to satisfy (iii).

As Jarczyk and Matkowski point out, it is easy to verify that, for $h : (0, \infty) \to (0, \infty)$, $\log \circ h \circ \exp$ is convex if and only if

$$h(\sqrt{xy}) \leq \sqrt{h(x)h(y)}, \text{ for all } x, y \text{ in } (0, \infty),$$

i.e., if and only if h is convex with respect to the geometric mean. Accordingly, following the terminology introduced in [Matkowski (1992)], any such function is said to be **geometrically convex**.

Tardiff [1984] also established the next two theorems.

Theorem 4.2.8 *Suppose that h_1 and h_2 both satisfy Condition (iii) and that the third derivative h_2''' exists. Then $h_1 \circ h_2$ satisfies Condition (iii) and, a fortiori, the inequality (4.2.4).*

Theorem 4.2.9 *Let ϕ be any continuous, non-decreasing, convex function such that*

$$\int_0^\infty e^{\phi(\log u)} du = \infty.$$

Let T_g be the strict t-norm additively generated by g and let

$$f(x) = \int_0^{g(x)} e^{\phi(\log u)}\, du.$$

Then f is an additive generator of a strict t-norm T_f which dominates T_g.

It follows from Theorem 4.2.2 that the dominance relation is reflexive and antisymmetric on the set $\mathcal{T}$ of all t-norms. This observation leads naturally to the question (see [Schweizer and Sklar (1983), (2005), Problem 12.11.3]): Is the relation "dominates" transitive, i.e., is $(\mathcal{T}, >>)$ a partially ordered set? If not, under what conditions, i.e., on which subsets of $\mathcal{T}$, is it transitive? To date, there are several partial answers.

Theorem 4.2.10 [Sherwood (1984)] *The dominance relation is transitive on the family of t-norms (2.6.1). Specifically, if for any α in $(-\infty, \infty)\backslash\{0\}$,*

$$T_\alpha(x, y) = [\max(x^{-\alpha} + y^{-\alpha} - 1, 0)]^{-1/\alpha}, \tag{4.2.6}$$

and if $T_0 = P$, then $T_\alpha >> T_\beta$ if and only if $\alpha \geq \beta$.

Since $T_\alpha \geq T_\beta$ if and only if $\alpha \geq \beta$, dominance and the usual pointwise order coincide in this family.

Lemma 4.2.11 *Let T_1, T_2 and T_3 be strict t-norms with additive generators t_1, t_2 and t_3, respectively, and let $h_1 = t_1 \circ t_2^{-1}$, $h_2 = t_2 \circ t_3^{-1}$ and $h_3 = t_1 \circ t_3^{-1}$. Suppose that $T_1 >> T_2$, $T_2 >> T_3$ and that $\log \circ h_3 \circ \exp$ is convex. Then $T_1 >> T_3$.*

Proof. Clearly h_3 is continuous and strictly increasing, with $h_3(0) = 0$. Since $T_1 \geq T_2$ and $T_2 \geq T_3$ we have $T_1 \geq T_3$ whence, by Theorem 2.2.2, h_3 is convex. Thus, since $\log \circ h_3 \circ \exp$ is convex by hypothesis, it follows from Mulholland's Condition (i) that $T_1 >> T_3$. □

Theorem 4.2.12 *The dominance relation is transitive on any subset $\mathcal{D}$ of $\mathcal{T}_{St}$ having the property that the additive generators t_1 and t_2 of every T_1 and T_2 in $\mathcal{D}$ are such that the function $h = t_1 \circ t_2^{-1}$ satisfies Condition (i).*

Using a corresponding argument, but based on Theorem 4.2.8, Tardiff [1984] established

Theorem 4.2.13 *The dominance relation is transitive on any subset $\mathcal{D}'$ of $\mathcal{T}_{St}$ having the property that the additive generators t_1 and t_2 of every T_1 and T_2 in $\mathcal{D}'$ are such that the function $h = t_1 \circ t_2^{-1}$ has a third derivative and satisfies Condition (iii).*

To illustrate, the t-norm T_α displayed in (4.2.6) is generated by the function t_α given by $t_\alpha(x) = (x^{-\alpha} - 1)/\alpha$ for $\alpha \neq 0$. Some simple calculations show that, for $\alpha, \beta > 0$, the function $h = t_\alpha \circ t_\beta^{-1}$ has a third derivative and satisfies Condition (iii). Thus – with $T_0 = P$ – Theorem 4.2.13 yields Theorem 4.2.10 whenever the t-norms T_α and T_β are strict. It does not yield the entire theorem since T_α is not strict when $\alpha < 0$.

Example 4.2.6 shows that the convexity of $t_1 \circ t_2^{-1}$ does not guarantee that the inequality (4.2.4) holds. However, Lemma 3 of [Tardiff (1980)] states that if $t_1 \circ t_2^{-1}$ is convex then the weaker form of (4.2.4) which is obtained by setting $v_2 = 0$ does hold. This fact, combined with Theorem 4.2.5 and the note in its proof, yields:

Theorem 4.2.14 *Let T_1 and T_2 be strict t-norms with additive generators t_1 and t_2, respectively, and let $h = t_1 \circ t_2^{-1}$. Then the following are equivalent:*

(a) $t_1 \circ t_2^{-1}$ *is convex;*

(b) $h^{-1}(h(u_1 + v_1) + h(u_2)) \leq h^{-1}(h(u_1) + h(u_2)) + v_1;$ (4.2.7)

(c) $T_1(T_2(x_1, y_1), x_2) \geq T_2(T_1(x_1, x_2), y_1).$ (4.2.8)

The inequality (4.2.8) motivates the following

Definition 4.2.15 Let $(S, \leq)$ be a partially ordered set and let A and B be binary operations on S. Then A **weakly dominates** B if, for all x_1, x_2, y_1 in S

$$A(B(x_1, y_1), x_2) \geq B(A(x_1, x_2), y_1). \tag{4.2.9}$$

It follows readily that Theorem 4.2.2 remains valid when "dominance" is replaced by "weak dominance", and again, Example 4.2.3 shows that, in general, weak dominance is not transitive, since $F(H(1,1),1) = 2$ while $H(F(1,1),1) = 1$.

Note that, in view of Theorem 4.2.14, the conclusion of Corollary 2.2.3 can be strengthened from pointwise order to weak dominance.

For t-norms, dominance clearly implies weak dominance, and Example 4.2.6 shows that the converse is false. Similarly if T_1 weakly dominates T_2, then $T_1 \geq T_2$; and the example following Corollary 2.2.4 shows that the converse is false. Thus these three partial order relations are distinct. Finally, since the composition of two increasing convex functions is increasing and convex, we have

Theorem 4.2.16 *The weak dominance relation is transitive on the set* $\mathcal{T}_{St}$ *of all strict t-norms.*

In conclusion, we note that a class of partial order relations which inlcudes the usual pointwise ordering and weak dominance is studied in [Alsina and Gimenez (1984)].

4.3 Uniformly close associative functions

In this section we present results relating the distance (in the sup metric) between Archimedean functions, to the distances between their additive generators. These results, some of which we extend, are due to [Alsina and Ger (1985), (1988)].

Definition 4.3.1 A pair of real-valued functions φ and ψ defined on a set A are (uniformly) **ϵ-close** for some $\epsilon > 0$ if

$$|\varphi(a) - \psi(a)| \leq \epsilon, \text{ for all } a \text{ in } A,$$

i.e., if

$$\|\varphi - \psi\|_A \leq \epsilon,$$

where $\|\varphi - \psi\|_A = \sup_{a \text{ in } A} |\varphi(a) - \psi(a)|$.

Note that this definition encompasses both one- and multi-place functions.

For a given associative function H_0 defined on an interval J and a given $\epsilon > 0$, we seek an associative function H on J such that H and H_0 are ϵ-close. The existence of such an H and its construction depend on the boundedness of J as well as on the boundary conditions and regularity assumptions imposed on H_0. The analysis in the Archimedean case is based in part on the following fundamental result on the "stability" of Cauchy's equation (1.2.1) due to D.H. Hyers [1941].

Lemma 4.3.2 *Suppose that φ is ϵ-additive on $\mathbb{R}^+$ (or $\mathbb{R}$) for some $\epsilon > 0$, i.e.,*

$$|\varphi(u+v) - \varphi(u) - \varphi(v)| \leq \epsilon, \text{ for all } u, v \text{ in } \mathbb{R}^+ (\text{or } \mathbb{R}). \tag{4.3.1}$$

Then there exists an additive function γ on $\mathbb{R}^+$ (or $\mathbb{R}$), i.e., a solution of Cauchy's functional equation (1.2.1), such that

$$\|\varphi - \gamma\| \leq \epsilon.$$

Moreover, if φ is continuous at a single point, then γ is continuous, whence $\gamma(u) = ku$.

The inequality (4.3.1) is often called the "Ulam-Hyers inequality". Note that if γ is additive and $\|\eta - \gamma\| \leq \epsilon$, an elementary application of the triangle inequality yields that for all u, v,

$$|\eta(u+v) - \eta(u) - \eta(v)| \leq 3\epsilon. \tag{4.3.2}$$

To illustrate, let G_0 be addition on $\mathbb{R}^+$ and suppose that G belongs to $\mathcal{G}_{St}(\mathbb{R}^+)$, with (continuous) additive generator g. Then

$$\|G - G_0\| \leq \epsilon$$

if and only if for all x, y in $\mathbb{R}^+$

$$|g^{-1}(g(x) + g(y)) - x - y| \leq \epsilon$$

if and only if for all u, v in $\mathbb{R}^+$

$$|g^{-1}(u+v) - g^{-1}(u) - g^{-1}(v)| \leq \epsilon,$$

i.e., if and only if g^{-1} is ϵ-additive.

Thus, if $\|G - G_0\| \leq \epsilon$, then by Lemma 4.3.2, for some $k > 0$, $|g^{-1}(x) - kx| \leq \epsilon$ for all x in $\mathbb{R}^+$, i.e., $\|g^{-1} - kj\| \leq \epsilon$. On the other hand, if $\|g^{-1} - kj\| \leq \epsilon$, then by (4.3.2), $\|G - G_0\| \leq 3\epsilon$.

The last inequality is in fact best-possible: for, on choosing g^{-1} so that $\|g^{-1} - j\| \leq \epsilon$ and $g^{-1}(x_0) = x_0 - \epsilon$ and $g^{-1}(2x_0) = 2x_0 + \epsilon$, for some $x_0 > \epsilon$, we have that $G(x_0 - \epsilon, x_0 - \epsilon) - G_0(x_0 - \epsilon, x_0 - \epsilon) = g^{-1}(2x_0) - (2x_0 - 2\epsilon) = 3\epsilon$.

We now begin to explore the relationship between closeness of continuous Archimedean functions and their generators. Before stating the first result, we observe that if an Archimedean function H on J has an additive generator h that satisfies the Lipschitz condition

$$|h(x) - h(y)| \leq m|x - y|, \text{ for all } x, y \text{ in } J, \tag{4.3.3}$$

then the generator $\frac{1}{m}h$ satisfies (4.3.3) with $m = 1$.

The Lipschitz condition (4.3.3) is very restrictive. It fails to hold, for example, when H belongs to $\mathcal{F}_{St}[a, b]$ and a is finite or when H belongs to $\mathcal{G}_{St}[a, b]$ and b is finite. When it does hold, as in the case of addition on $\mathbb{R}$ or $\mathbb{R}^+$, we have

Theorem 4.3.3 *Let H and H_0 belongs to $\mathcal{F}_{St}(J)$ or $\mathcal{G}_{St}(J)$ for some (infinite) interval J. Suppose that H_0 has an additive generator h_0 that satisfies the Lipschitz condition (4.3.3) with $m = 1$ and that $\|H - H_0\| \leq \epsilon$. Then there exists a generator h of H such that $\|h - h_0\|_J \leq \epsilon$, i.e., if H and H_0 are ϵ-close, then so are h and h_0.*

Proof. By hypothesis h_0 satisfies

$$|h_0(x) - h_0(y)| \leq |x - y|, \text{ for all } x, y \text{ in } J.$$

Suppose that $\|H - H_0\| \leq \epsilon$ and h_1 is a generator of H, and let $\varphi : \mathbb{R}^+ \to \mathbb{R}^+$ be the function defined by $\varphi = h_0 \circ h_1^{-1}$. Then for all u, v in $\mathbb{R}^+$,

$$\begin{aligned} |\varphi(u + v) - \varphi(u) - \varphi(v)| &\leq |h_0^{-1}[\varphi(u + v)] - h_0^{-1}[\varphi(u) + \varphi(v)]| \\ &= |h_1^{-1}(u + v) - h_0^{-1}[(h_0 \circ h_1^{-1})(u) + (h_0 \circ h_1^{-1})(v)]| \\ &= |H(h_1^{-1}(u), h_1^{-1}(v)) - H_0(h_1^{-1}(u), h_1^{-1}(v))| \\ &\leq \epsilon, \end{aligned}$$

whence by Lemma 4.3.2, $\|\varphi - kj\| \leq \epsilon$, or equivalently, letting $h = kh_1$, $\|h_0 - h\| \leq \epsilon$ for some $k > 0$. □

Theorem 4.3.4 *Let H_0, H belongs to $\mathcal{F}_{St}(J)$ or $\mathcal{G}_{St}(J)$ for some interval J, and suppose that for some additive generator h_0 of H_0, the inverse function h_0^{-1} satisfies the Lipschitz condition*

$$|h_0^{-1}(u) - h_0^{-1}(v)| \leq m|u - v|, \text{ for all } u, v \text{ in } \mathbb{R}^+. \tag{4.3.4}$$

If for some generator h of H, $\|h - h_0\|_J \leq \epsilon$ then $\|H - H_0\|_{J^2} \leq 3m\epsilon$.

Proof. If $\|h - h_0\| \le \epsilon$, then, with $\varphi = h_0 \circ h^{-1}$, for all u in $\mathbb{R}^+$, we have $|\varphi(u) - u| \le \epsilon$, whence for all x, y in $\mathbb{R}^+$,

$$\begin{aligned} |H(x,y) - H_0(x,y)| &\le m|h_0(H(x,y)) - h_0(H_0(x,y))| \\ &= m|\varphi[h(x) + h(y)] - \varphi(h(x)) - \varphi(h(y))| \\ &\le m[|\varphi(h(x) + h(y)) - (h(x) + h(y))| \\ &\quad + |\varphi(h(x)) - h(x)| + |\varphi(h(y)) - h(y)|] \\ &\le 3m\epsilon. \end{aligned}$$

□

The following important example shows that if h_0^{-1} fails to satisfy (4.3.4), then there need not be any associative function whatever that is close to H_0.

Example 4.3.5 Let $G : [1,\infty]^2 \to [1,\infty]$ be associative, and suppose that for some $\epsilon > 0$ $\|G - P_{[1,\infty]}\| \le \epsilon$, i.e.,

$$|G(x,y) - xy| \le \epsilon, \text{ for all } x, y \text{ in } [1,\infty].$$

Then $G = P_{[1,\infty]}$.

To verify this, fix x, y in $[1,\infty]$. The triangle inequality, together with the associativity of G yields that for every z in $[1,\infty)$,

$$\begin{aligned} |G(x,y)z - xyz| &\le |G(G(x,y),z) - G(x,y)z| \\ &\quad + |G(x,G(y,z)) - xG(y,z)| \\ &\quad + |xG(y,z) - xyz| \\ &\le \epsilon + \epsilon + x\epsilon, \end{aligned}$$

so that

$$|G(x,y) - xy| \le \left(\frac{2+x}{z}\right)\epsilon.$$

Since z may be arbitrarily large, it follows that $|G(x,y) - xy| = 0$, and so $G = P_{[1,\infty]}$ (For an extension of this example and some related results, see [Alsina (1991)].)

This result holds *a fortiori* on any larger interval, but not on I. For consider the t-norms T_α of Family (2.6.1). As $\alpha \to 0$, T_α converges uniformly to P.

We now turn to t-norms (and tacitly, with the obvious modifications, to s-norms).

For continuous Archimedean t-norms T_1 and T_2, with generators t_1 and t_2, respectively, there is in general no relationship between $\|T_1 - T_2\|$ and $\|t_1 - t_2\|$.

(a) If both T_1 and T_2 are non-strict then, for each $\epsilon > 0$, we may choose t_1 and t_2 such that $t_1(0) = t_2(0) = \epsilon$, whence $\|t_1 - t_2\| \leq \epsilon$, while there is no restriction on $\|T_1 - T_2\|$.

(b) If T_1 is strict and T_2 is not then $\|t_1 - t_2\| \geq |t_1(0) - t_2(0)| = \infty$, regardless of the value of $\|T_1 - T_2\|$.

(c) If T_1 and T_2 are both strict then, under its stated hypotheses, Theorem 4.3.4 can be applied. On the other hand, for Family (2.6.1), as $\alpha \to 0$, $T_\alpha \to P$ uniformly, yet for all positive a and b,

$$\lim_{x \searrow 0} \left| a\left(\frac{1}{x^\alpha} - 1\right) + b \log x \right| = \infty,$$

whence $\|at_\alpha - bt_0\| = \infty$.

Our last two results concern copulas.

Theorem 4.3.6 *Let T_0 be a strict copula with generator t_0, let φ be a continuous and strictly increasing function from $\mathbb{R}^+$ onto $\mathbb{R}^+$ that satisfies the Ulam-Hyers inequality (4.3.1), and let T be the t-norm generated by $t = \varphi^{-1} \circ t_0$. Then $\|T - T_0\| \leq 1 - t_0^{-1}(\epsilon)$.*

Proof. The convexity of t_0^{-1} immediately yields that

$$|t_0^{-1}(u) - t_0^{-1}(v)| \leq 1 - t_0^{-1}(|u - v|), \text{ for all } u, v \text{ in } \mathbb{R}^+.$$

Hence, putting $u = t(x)$ and $v = t(y)$, we obtain

$$\begin{aligned} |T(x,y) - T_0(x,y)| &= \left|(t_0^{-1} \circ \varphi)(u+v) - t_0^{-1}[\varphi(u) + \varphi(v)]\right| \\ &\leq 1 - t_0^{-1}(|\varphi(u+v) - \varphi(u) - \varphi(v)|) \\ &\leq 1 - t_0^{-1}(\epsilon). \end{aligned}$$

□

Such φ are easily constructed: let $\alpha : \mathbb{R}^+ \to \left[0, \frac{1}{3}\epsilon\right]$ be continuous and strictly increasing, with $\alpha(0) = 0$, and let $\varphi(u) = u + \alpha(u)$. Note that when φ is concave (α concave), then $\phi^{-1} \circ t_0$ is convex so that T is also a copula.

Theorem 4.3.7 *Suppose that T_1, T_2 are Archimedean copulas having generators t_1, t_2 that satisfy $\|t_1 - t_2\| \leq \epsilon$. Then $\|T_1 - T_2\| \leq 3\epsilon$.*

Proof. Since $\|T_1 - T_2\| \le \|\text{Min} - W\| = \frac{1}{2}$, the conclusion holds trivially when $\epsilon \ge 1/6$, so we may assume that $\epsilon < 1/6$. We may confine our attention to (x, y) in $[3\epsilon, 1 - 3\epsilon]^2$ since elsewhere in I^2 we know that $|T_1(x, y) - T_2(x, y)| \le \text{Min}(x, y) - W(x, y) \le 3\epsilon$.

By hypothesis, $t_2^{(-1)}(u) - \epsilon \le t_1^{(-1)}(u) \le t_2^{(-1)}(u) + \epsilon$ for all $u \ge 0$, which implies that $t_2^{(-1)}(t_1(x)) \le x + \epsilon$, and thus that $t_1(x) \ge t_2(x + \epsilon)$. Therefore,

$$\begin{aligned} T_1(x,y) &= t_1^{(-1)}[t_1(x) + t_1(y)] \le t_2^{(-1)}[t_1(x) + t_1(y)] + \epsilon \\ &\le t_2^{(-1)}[t_2(x+\epsilon) + t_2(y+\epsilon)] + \epsilon \\ &= T_2(x+\epsilon, y+\epsilon) + \epsilon \\ &\le T_2(x,y) + 3\epsilon, \end{aligned}$$

where the last inequality holds by virtue of (1.4.2). Upon interchanging the rolls of t_1 and t_2 we obtain $T_2(x, y) \le T_1(x, y) + 3\epsilon$, which completes the argument. □

To conclude this section we mention an interesting (and almost surely difficult) open problem suggested by Hyers' Theorem: Is the associativity equation stable? More precisely:

A function $F : J^2 \to J$ is said to be **ϵ-associative** if

$$|F(F(x,y),z) - F(x,F(y,z))| \le \epsilon, \text{ for all } x, y, z \text{ in } J.$$

If F is ϵ-associative, is there necessarily an associative function A on J such that

$$\|F - A\|_J \le m\epsilon$$

for some $m > 0$? For $m = 1$?

4.4 Serial iterates and n-copulas

For $n = 2, 3, \ldots$, the **serial iterates** of a t-norm T are the functions $T^n : I^n \to I$ defined recursively via

$$T^2 = T,$$

and

$$T^{n+1}(x_1, \ldots, x_{n+1}) = T(T^n(x_1, \ldots, x_n), x_{n+1}). \tag{4.4.1}$$

By virtue of the associativity and the commutativity of T, each T^n is symmetric in all places. Note that the boundary conditions (1.3.1) yield that

$$T^n(0, x_2, ..., x_n) = 0, \tag{4.4.2a}$$

$$T^n(x, 1, 1, ..., 1) = x, \tag{4.4.2b}$$

$$T^{n+1}(x_1, ..., x_n, 1) = T^n(x_1, ..., x_n), \tag{4.4.2c}$$

and so forth (by symmetry). Moreover, each T^n is non-decreasing in each place.

For the serial iterates of a continuous Archimedean t-norm, we have:

Lemma 4.4.1 *If T is the continuous Archimedean t-norm generated by t, then for each $n \geq 2$,*

$$T^n(x_1, ..., x_n) = t^{(-1)}(t(x_1) + \cdots + t(x_n)). \tag{4.4.3}$$

Proof. For $n = 3$ we have

$$\begin{aligned} T^3(x, y, z) &= T(T(x, y), z) \\ &= t^{(-1)}(t \circ t^{(-1)}(t(x) + t(y)) + t(z)) \\ &= t^{(-1)}(\min[t(x) + t(y), t(0)] + t(z)) \\ &= t^{(-1)}(\min[t(x) + t(y) + t(z), t(0) + t(z)]) \\ &= t^{(-1)}(t(x) + t(y) + t(z)), \end{aligned}$$

regardless of whether or not $t(x) + t(y) \geq t(0)$. The result then follows from an induction argument that mirrors the above. □

Note in particular that

$$W^n(x_1, ..., x_n) = \max(x_1 + \cdots + x_n - n + 1, 0),$$

$$P^n(x_1, ..., x_n) = x_1 \cdots x_n,$$

and

$$\mathrm{Min}^n(x_1, ..., x_n) = \mathrm{Min}(x_1, ..., x_n).$$

For any sequence $\{x_n\}$, x_n in I, and any t-norm T, the sequence $\{T^n(x_1, ..., x_n)\}$ is non-increasing and thus has a limit in I. Accordingly, we may define

$$T^{\infty}(\{x_n\}) = \lim_{n \to \infty} T^n(x_1, ..., x_n).$$

Clearly, if T is a continuous Archimedean t-norm with generator t, then $T^{\infty}(\{x_n\}) > 0$ if and only if $\sum_{n=1}^{\infty} t(x_n) < t(0)$. Thus, for strict t-norms, convergence of the infinite T-product $T^{\infty}(\{x_n\})$ is equivalent to convergence of the infinite series $\sum_{n=1}^{\infty} t(x_n)$. (See Section 1.8 of [Hadžić and Pap (2001)] for further details and examples.)

Many of our results on continuous Archimedean t-norms can be carried over to their serial iterates. As an illustration, consider the extension of a diagonal to t-norms provided in Theorem 3.8.2 (and the discussion preceding it). Let δ satisfy the conditions of Definition 3.8.1. Then for any fixed integer $n \geq 2$, there exist continuous Archimedean t-norms T such that

$$T^n(x, x, ..., x) = \delta(x), \text{ for all } x \text{ in } I.$$

The proof involves replacing equation (3.8.3) by the Schröder equation

$$t(\delta(x)) = nt(x), \text{ for } d \leq x \leq 1,$$

whose solutions generate all such T. Moreover, t is unique when $\delta'(1-) = n$ (see [Kimberling (1973a)]).

Our principal interest is to determine when the serial iterates of an Archimedean copula are higher dimensional copulas – the class of functions that extend Theorem 1.4.3 from bivariate to multivariate distributions. We begin with a brief description of this class; for complete details, including proofs, consult [Nelsen (1999); Schweizer and Sklar (1983), (2005)].

An **n-box** is a subset B of I^n of the form

$$B = J_1 \times \cdots \times J_n, \quad J_k = [a_k, b_k] \subseteq I.$$

For points $\vec{x} = (x_1, ..., x_n)$ and $\vec{y} = (y_1, ..., y_n)$ in I^n, write $\vec{x} \leq \vec{y}$ (resp., $\vec{x} < \vec{y}$) if $x_k \leq y_k$ (resp., $x_k < y_k$) for $k = 1, 2, ..., n$. Thus, $B = \{\vec{z} \text{ in } I^n : \vec{a} \leq \vec{z} \leq \vec{b}\}$ and we write $B = [\vec{a}, \vec{b}]$. In particular, $I^n = [\vec{0}, \vec{1}]$. The **vertices** of B are the points $\vec{v} = (v_1, ..., v_n)$ such that each v_k equals either a_k or b_k. If $\vec{a} < \vec{b}$, then for each vertex $\vec{v}$ of B, define

$$\operatorname{sgn}_B(\vec{v}) = \begin{cases} 1, & \text{if } v_k = a_k \text{ for an even number of } k\text{'s}, \\ -1, & \text{if } v_k = a_k \text{ for an odd number of } k\text{'s}; \end{cases}$$

otherwise, let $\operatorname{sgn}_B(\vec{v}) = 0$.

Now let $C : I^n \to I$ be a given function. For each n-box B define the **C-volume of B** via

$$V_C(B) = \sum_{\vec{v} \text{ in } B} \operatorname{sgn}_B(\vec{v}) C(\vec{v}). \tag{4.4.4}$$

(The right-hand side is just an elegant way of writing the n-th order iterated difference of C over B). We say that C is **n-increasing** if $V_C(B) \geq 0$ for all n-boxes B in I^n.

Definition 4.4.2 An **n-dimensional copula** (briefly, **n-copula**) is an n-place function $C : I^n \to I$ that satisfies the following conditions:

(i) On the boundary of I^n,

$$C(\vec{x}) = 0 \text{ if at least one coordinate of } \vec{x} \text{ is } 0, \tag{4.4.5a}$$

$$C(\vec{x}) = x_k \text{ if all coordinates of } \vec{x} \text{ are 1 except possibly } x_k. \tag{4.4.5b}$$

(ii) C is n-increasing, i.e.,

$$V_C[\vec{x}, \vec{y}] \geq 0, \text{ whenever } \vec{0} \leq \vec{x} \leq \vec{y} \leq \vec{1}. \tag{4.4.6}$$

Note that this extends Definition 1.4.1, since when $n = 2$,

$$V_C[\vec{x}, \vec{y}] = C(x_2, y_2) - C(x_2, y_1) - C(x_1, y_2) + C(x_1, y_1).$$

Some key basic properties of n-copulas are collected in

Lemma 4.4.3 *Let C be an n-copula for some $n \geq 3$. Then:*
(a) $0 \leq C(\vec{y}) - C(\vec{x}) \leq \sum_{k=1}^{n} (y_k - x_k)$, *whenever* $\vec{0} \leq \vec{x} \leq \vec{y} \leq \vec{1}$, (4.4.7)
whence C is non-decreasing in each place and Lipschitz continuous.

(b) For $2 \leq m < n$, the m-margins of C are m-copulas. (For $m \geq 1$ the **m-margins** *are the $\binom{n}{m}$ functions from I^m to I obtained from C by setting $n - m$ of its arguments equal to 1.)*

For a proof of part (a), which is the extension of the first part of Lemma 1.4.2, as well as part (b), see [Schweizer and Sklar (1983), (2005)].

We can now state the n-dimensional version of Theorem 1.4.3.

Theorem 4.4.4 *Let H be an n-dimensional probability distribution with margins $F_1, ..., F_n$. Then there is an n-copula C such that for all $\vec{u}$ in $\mathbb{R}^n$,*

$$H(\vec{u}) = C(F_1(u_1), ..., F_n(u_n)). \tag{4.4.8}$$

If $F_1, ..., F_n$ are continuous, then C is unique; otherwise C is uniquely determined on $(\text{Ran } F_1) \times \cdots \times (\text{Ran } F_n)$.

In the other direction, for any univariate distributions $F_1, ..., F_n$ and any copula C, the function H defined by (4.4.8) is an n-dimensional probability distribution with 1-margins $F_1, ..., F_n$.

It is easy to show that Min^n and P^n are n-copulas for all n: in fact,

$$V_{\text{Min}^n}[\vec{x}, \vec{y}] = \max(\min y_k - \max x_k, 0)$$

and

$$V_{P^n}[\vec{x}, \vec{y}] = \prod_{k=1}^{n} (y_k - x_k).$$

Min^n is the copula of increasing dependence, and P^n the copula of mutual independence; see Theorem 1.4.4, parts (a) and (b). On the other hand, W^n is not a copula for $n \geq 3$ since an easy calculation yields

$$V_{W^n}[\vec{1/2}, \vec{1}] = 1 - \frac{n}{2}.$$

However, the counterpart of the Fréchet-Hoeffding bounds (1.4.3) holds, viz, for any n-copula C,

$$W^n \leq C \leq M^n.$$

For details, see [Schweizer and Sklar (1983), (2005)]; and for historical motivation, see especially [Fréchet (1935), (1951); Sklar (1959); Dall'Aglio (1960), (1972); Hailperin (1965)].

We next explore the connection between copulas and serial iterates of Archimedean t-norms. Since the boundary conditions (4.4.5) and (4.4.2a,b) agree, any such T^n is an n-copula if and only if it is n-increasing. Moreover, by virtue of Lemma 4.4.3(b) with $m = 2$, T must be a 2-copula. Thus the problem is the following: Given an Archimedean 2-copula C, find conditions on its generator t so that C^n is n-increasing. It will often be convenient in the sequel to denote the outer generator $t^{(-1)}$ of C by g, in which case

$$C(x, y) = g(t(x) + t(y)),$$

where $g : \mathbb{R}^+ \to I$ is continuous, strictly decreasing, and convex on $[0, t(0)]$, with $g(0) = 1$ and $g(t(0)) = 0$; and by (4.4.3),

$$C^n(\vec{x}) = g(t(x_1) + \cdots + t(x_n)) \tag{4.4.9}$$

for all $\vec{x} = (x_1, ..., x_n)$ in I^n. Such a generator t will then provide a source of n-dimensional probability distributions.

Definition 4.4.5 A real function g is **n-monotonic** on an interval J if it is n times differentiable and for all x in J,

$$(-1)^k g^{(k)}(x) \geq 0, \text{ for } 0 \leq k \leq n; \tag{4.4.10a}$$

and g is **completely monotonic** (c.m.) if it is n-monotonic for all n, i.e., if

$$(-1)^k g^{(k)} \geq 0, \text{ for all } k \geq 0. \tag{4.4.10b}$$

Note that if g is 2-monotonic then g is positive, decreasing, and convex, and that if g is n-monotonic then it is m-monotonic for each $m < n$.

The class of c.m. functions has been studied extensively; see [Widder (1941); Boas (1971)]. Key properties are the following:

(1) $g : \mathbb{R}^+ \to I$, with $g(0) = 1$, is c.m. on $(0, \infty)$ if and only if it is (real) analytic, and

$$g(u) = \int_0^\infty e^{-ux} dG(x) \tag{4.4.11}$$

for some G in Δ^+, i.e., g is the Laplace transform of a probability distribution G for which $G(0) = 0$.

(2) g is c.m. on $(0, \infty)$ if and only if g is non-negative and

$$\sum_{m=0}^{n} (-1)^{n-m} \binom{n}{m} g(u - mh) \geq 0 \tag{4.4.12}$$

for all integers $n \geq 1$ and all u and h such that

$$0 \leq u - nh < \cdots < u - h < u < \infty.$$

The following basic result is due to A. Sklar [1973] (see also [Kimberling (1974)]).

Theorem 4.4.6 *Let C be a strict copula with additive generator t. Then the serial iterates C^n of C are n-copulas for all $n \geq 2$ if and only if g $(= t^{-1})$ is completely monotonic on $(0, \infty)$. Furthermore, in this case $C \geq P$.*

Proof. Suppose that C^n is an n-copula for all $n \geq 2$, and consider the n-box $B = [\vec{a}, \vec{b}]$, where $\vec{a} = (a, ..., a)$ and $\vec{b} = (b, ..., b)$ are points on the main diagonal of I^n with $a < b$, so that $\vec{a} < \vec{b}$. Since C is commutative and associative, it follows that, for any vertex $\vec{v}$ of B, $C^n(\vec{v})$ depends only on the number m of coordinates of $\vec{v}$ that are equal to a. Thus (4.4.4) and (4.4.9) yield

$$\begin{aligned} 0 \leq V_{C^n}(B) &= \sum_{m=0}^{n} (-1)^m \binom{n}{m} g(mt(a) + (n-m)t(b)) \\ &= \sum_{m=0}^{n} (-1)^m \binom{n}{m} g(nt(a) - (n-m)(t(a) - t(b))) \\ &= \sum_{m=0}^{n} (-1)^{n-m} \binom{n}{m} g(nt(a) - m(t(a) - t(b))). \end{aligned}$$

Now, for any u in $(0, \infty)$ and any h such that $h < u/n$, choose a, b in I so that $t(a) = u/n$ and $t(a) - t(b) = h$. Then the last summation above reduces to (4.4.12), whence g is completely monotonic.

For the converse, suppose that the outer generator g is completely monotonic on $(0, \infty)$. Since g is strictly decreasing and convex, it follows easily that $-\infty < t'(x) < 0$ for all x in (0,1). Consequently, for each n, the n'th order partial derivative $D_{12\cdots n}C^n$ exists on $(0,1)^n$ and is given by

$$D_{12\cdots n}C^n(x_1, ..., x_n) = t'(x_1)\cdots t'(x_n)g^{(n)}(t(x_1) + \cdots + t(x_n)),$$

which is non-negative since $t'(x_1)\cdots t'(x_n)$ and $g^{(n)}$ have the same parity whenever $g^{(n)}(t(x_1) + \cdots + t(x_n))$ is non-zero. The integral of this partial derivative over any n-box in I^n is therefore non-negative, so that C^n is n-increasing and thus an n-copula.

Finally, if g is c.m., then by (4.4.11), for some G in Δ^+

$$\begin{aligned} t^{-1}(u+v) &= \int_0^\infty e^{-ux}e^{-vx}dG(x) \\ &\geq \int_0^\infty e^{-ux}dG(x) \int_0^\infty e^{-vx}dG(x) \\ &= t^{-1}(u)t^{-1}(v). \end{aligned}$$

(The inequality is a consequence of Problem 7.17 in [Apostol (1974)].) Now upon setting $u = t(x)$ and $v = t(y)$, we obtain $C(x, y) \geq xy$. □

The proof of Theorem 4.4.6 immediately yields

Corollary 4.4.7 *If g is n-monotonic then C^n is an n-copula.*

For a given strict copula $C > P$, it may be difficult to determine directly whether t^{-1} is c.m. and thus whether C^n is a copula for all n. In fact, this is the situation for most of the families of copulas in Table 2.6. Here the following three criteria are highly effective:

(1) If φ is c.m. and ψ is absolutely monotonic (i.e., $\psi^{(k)} \geq 0$ for all $k \geq 0$), then $\psi \circ \varphi$ is c.m.
(2) If φ is c.m. and ψ is a positive function such that ψ' is c.m., then $\varphi \circ \psi$ is c.m.
(3) If φ and ψ are c.m., then so is their product $\varphi \cdot \psi$.

(Note that the monotonicity conditions must be enforced on appropriate intervals.) The first of these criteria is given in [Widder (1941)] and the last two in [Feller (1966), §XIII. 4]. The proofs involve straightforward induction.

Among the families of strict copulas T_α in Table 2.6, those for which t_α^{-1} is indeed c.m. throughout the range of parameter values for which $T_\alpha \geq P$ are Families 1, 3, 4, 5, 6, 12, 15, 16, 22 and 23. Verification for several of these is presented in [Nelsen (1999)].

While the condition $C \geq P$ in Theorem 4.4.6 is necessary, it is not sufficient:

Example 4.4.8
(a) For Family (2.6.18), $T_\alpha > P$ whenever $\alpha \geq 1$, but t_α^{-1} is not c.m. However, t_α^{-1} is 3-monotonic so that T_α^3 is a 3-copula. In fact, there is a sequence $\alpha_3 < \alpha_4 < \cdots$, with $\lim \alpha_n = \infty$, such that T_α^n is an n-copula if and only if $\alpha \geq \alpha_n$. This sequence begins $\alpha_3 = 1$, $\alpha_4 = \frac{1}{2}(3 + \sqrt{5})$, $\alpha_5 = \frac{1}{2}(5 + \sqrt{21})$. See [Frank (2002)].

(b) Let C be the strict copula generated by $t(x) = x^{-2} - x^2$. An application of Corollary 2.2.3 yields that $C > P$, but C^3 is not a 3-copula; for with $g(u) = t^{-1}(u) = \left[\frac{1}{2}(u^2 + 1)^{1/2} - \frac{1}{2}u\right]^{1/2}$, a tedious calculation gives $g'''(u) > 0$ when $u < 1/\sqrt{6}$.

The representation (4.4.11) of a c.m. function has been exploited in [Marshall and Olkin (1988)] to construct an extensive variety of families of (generators of) multivariate distributions. One inherent limitation of this approach, however, is that necessarily $C^n \geq P^n$ for all $n \geq 2$. We cite one example: for Family (2.6.3) with $0 \leq \alpha \leq 1$,

$$t_\alpha^{-1}(u) = (1-\alpha)/(e^u - \alpha) = \int_0^\infty e^{-ux} dG_\alpha(x),$$

where $G_\alpha(x) = (1-\alpha) \sum_{0 \leq k < x} \alpha^k$ for $x > 0$.

When t^{-1} is *not* c.m. – in particular, when $C < P$ – it is often a non-trivial problem to find the largest value of n for which C^n is an n-copula. Suppose, for instance, that a family of strict copulas $C_\alpha < P$ is positively directed and that $\lim_{\alpha \nearrow \gamma} C_\alpha = P$. Then it is natural – and consistent with the preceding discussion – that the answer take the following form: *There is a sequence $\alpha_2 < \alpha_3 < \cdots < \gamma$, with $\lim \alpha_n = \gamma$, such that C_α^n is an n-copula if and only if $\alpha \geq \alpha_n$.*

The existence of such sequences has been confirmed for Families (2.6.3), (2.6.5), (2.6.9), and (2.6.15) ($\gamma = 0, 0, \infty$, and 1, resp.) via an indirect method involving analysis of the roots of certain sequences of polynomials [Frank (2002)]. For example, for Family (2.6.3), $-1 \leq \alpha < 0$, it turns out that $\alpha_n = -1/r_n$, where r_n is the largest root of the polynomial P_n of degree $n-1$ in the sequence defined recursively by

$$P_2(x) = x - 1, \quad P_{n+1}(x) = (nx-1)P_n(x) - x(x+1)P_n'(x).$$

The first few values are $\alpha_2 = -1$, $\alpha_3 = \sqrt{3} - 2$, $\alpha_4 = 2\sqrt{6} - 5$. Thus,

$$C_\alpha^3(x,y,z) = xyz/[1 - \alpha(2 - x - y - z + xyz) + \alpha^2(1-x)(1-y)(1-z)]$$

is a 3-copula if and only if $3 - \sqrt{2} \leq \alpha \leq 1$.

For non-strict Archimedean copulas, there is a result parallel to Corollary 4.4.7, with slightly modified hypotheses on $g = t^{(-1)}$. When applied to Family (2.6.1), $-1 < \alpha < 0$, a simple, direct calculation yields that $\alpha_n = \frac{-1}{n-1}$, i.e., that C_α^n is an n-copula if $\alpha > \frac{-1}{n-1}$.

Multivariate copulas obtained as serial iterates of Archimedean copulas have proved useful in a number of statistical studies and in related applications, e.g., [Jouini and Clemen (1996); Marco and Ruiz-Rívas (1992); Meester and MacKay (1994)].

If C is an associative copula then, for any n and each $m < n$, the $\binom{n}{m}$ m-margins of the serial iterate C^n are identical. From this and the symmetry of all the serial iterates it follows that if $\{X_k\}$ is a sequence of continuous random variables such that C^n is the copula of any subset of size n, then the random variables $X_1, ..., X_k, ...$ are exchangeable in the sense of di Finetti [Chow and Teicher (1997)].

4.5 Positivity

We conclude this chapter with some results concerning notions of positivity for t-norms and relate these results to the preceding sections. The primary source is [Alsina and Tomás (1992)].

Definition 4.5.1 A function $H : [a,b]^2 \to \mathbb{R}$ is said to be **totally positive of order k** (briefly, TP_k) if, for each positive integer $m \leq k$ and all $a \leq x_1 \leq x_2 \leq \cdots \leq x_m \leq b$ and $a \leq y_1 \leq y_2 \leq \cdots \leq y_m \leq b$, the order m determinant $\det H(x_i, y_i)$ satisfies

$$\det H(x_i, y_j) \geq 0; \tag{4.5.1}$$

it is **positive semidefinite of order k** (briefly, SP_k) if (4.5.1) holds when $x_i = y_i$ for $i = 1, 2, ..., m$, i.e., if for each $m \leq k$

$$\det H(x_i, x_j) \geq 0 \tag{4.5.2}$$

whenever $a \leq x_1 \leq x_2 \leq \cdots \leq x_m \leq b$.

Total positivity – and the weaker positive semidefinite condition – play a major role in a number of areas of analysis and in statistics [Karlin (1968); Marshall and Olkin (1979), (1988)].

Note that when H is commutative the matrix $(H(x_i, x_j))$ is symmetric, so that SP_k means positive semidefinite in the usual algebraic sense.

Let $k = 2$. Then H is TP_2 if and only if H is non-negative and

$$H(x_1, y_1)H(x_2, y_2) - H(x_1, y_2)H(x_2, y_1) \geq 0 \tag{4.5.3}$$

whenever $x_1 \leq x_2$ and $y_1 \leq y_2$; and H is SP_2 if and only if

$$H(x,x)H(y,y) - H(x,y)H(y,x) \geq 0, \text{ for all } x, y. \tag{4.5.4}$$

Observe that if $T : I^2 \to I$ satisfies the t-norm boundary conditions (1.3.1), then T is a copula if and only if the function $H = \exp T$ is TP_2.

We now investigate the conditions TP_2 and SP_2 among t-norms.

Lemma 4.5.2 *Let $T : I^2 \to I$ satisfy (1.3.1).*

(a) If T is TP_2, then T weakly dominates P, whence $T \geq P$.
(b) If T is a continuous t-norm and if T weakly dominates P, then T is SP_2.
(c) If T is a continuous Archimedean t-norm with generator t, then T is SP_2 if and only if the function $\varphi(u) = t(e^{-u})$ is convex, in which case T weakly dominates P and T is a copula.

Proof. (a) For all x, y, z in (0,1), (4.5.3) yields that

$$T(x,yz)y = T(x,yz)T(1,y) \geq T(x,y)T(1,yz) = T(x,y)yz,$$

whence $T(x,yz) \geq T(x,y)z$, which is (4.2.9) with $A = T$ and $B = P$.

(b) Suppose that T weakly dominates P. Fix arbitrary x, y in I with $x < y$, choose z in I so that $x = T(y,z)$, and let $w = T(x,y)/T(y,y)$. Then

$$\begin{aligned} T(x,x)T(y,y) &= T(x,T(y,z))T(y,y) \\ &= T(T(x,y),z)T(y,y) \\ &= T(wT(y,y),z)T(y,y) \\ &\geq T(T(y,y),z)wT(y,y) \\ &= T(y,T(y,z))T(x,y) \\ &= [T(x,y)]^2, \end{aligned}$$

whence T satisfies (4.5.4).

(c) First note that if T is SP_2, then

$$T(x,x) = T(x,x)T(1,1) \geq [T(x,1)]^2 = x^2,$$

so that T must be strict. Choose arbitrary u, v in $\mathbb{R}^+$ and set $x = t^{-1}(u/2)$, $y = t^{-1}(v/2)$. Then since $T(x,y) = t^{-1}(\frac{u+v}{2})$ and $T(x,x)T(y,y) = t^{-1}(u)t^{-1}(v)$, it follows that T is SP_2 if and only if

$$t^{-1}(u)t^{-1}(v) \geq \left[t^{-1}\left(\frac{u+v}{2}\right)\right]^2, \text{ for all } u, v \text{ in } \mathbb{R}^+. \tag{4.5.5}$$

Let $\varphi(u) = t(e^{-u})$, so that $\varphi^{-1}(w) = -\log t^{-1}(w)$. Then, upon taking logarithms, (4.5.5) is equivalent to

$$\varphi^{-1}(u) + \varphi^{-1}(v) \leq 2\varphi^{-1}\left(\frac{u+v}{2}\right),$$

i.e., to the concavity of φ^{-1} and, equivalently, to the convexity of φ. Now by Theorem 4.2.14, T weakly dominates P; moreover, since $t(x) = \varphi(-\log x)$, t is convex and hence T is a copula. □

Theorem 4.5.3 *Let T be a continuous Archimedean t-norm. If T is SP_2 then T is TP_2. Thus, within the set $\mathcal{T}_{Ar}$, the conditions TP_2, SP_2, and weak dominance of P are equivalent.*

Proof. In view of Lemma 4.5.2, it suffices to show that SP_2 implies TP_2 for strict T.

Choose $x_1 < x_2$ and $y_1 < y_2$, all in I; set $u = t(x_1) + t(y_1)$, $v = t(x_2) + t(y_2)$, and $\lambda = [t(y_1) - t(y_2)]/[t(x_1) - t(x_2) + t(y_1) - t(y_2)]$. Then λ belongs to I,

$$\begin{aligned} t(x_2) + t(y_1) &= \lambda u + (1-\lambda)v, \\ t(x_1) + t(y_2) &= (1-\lambda)u + \lambda v, \end{aligned}$$

and, with the convex function φ given in Lemma 4.5.2(c), we have

$$\begin{aligned} T(x_1, y_1)T(x_2, y_2) &= t^{-1}(u)t^{-1}(v) = \exp[-\varphi^{-1}(u) - \varphi^{-1}(v)] \\ &= \exp\{-[\lambda\varphi^{-1}(u) + (1-\lambda)\varphi^{-1}(v)] \\ &\qquad -[(1-\lambda)\varphi^{-1}(u) + \lambda\varphi^{-1}(v)]\} \\ &\geq \exp\{-\varphi^{-1}[\lambda u + (1-\lambda)v] - \varphi^{-1}[(1-\lambda)u + \lambda v]\} \\ &= \exp\left[-\varphi^{-1}(t(x_2) + t(y_1)) - \varphi^{-1}(t(x_1) + t(y_2))\right] \\ &= T(x_2, y_1)T(x_1, y_2), \end{aligned}$$

whence T satisfies (4.5.3). □

Note that if t is differentiable, then since $xt'(x) = -\varphi'(-\log x)$, it follows that T is SP_2 if and only if $xt'(x)$ is increasing, in which case $t(x) = \int_1^x \frac{g(u)}{u} du$ for some increasing g. (Compare to Corollary 2.2.5.)

Corollary 4.5.4 *Let T be a strict t-norm with generator t. Then:*
(a) T is SP_2 if and only if

$$[T(x, y)]^2 \leq xT(x, T(y, y)), \quad \textit{for all } x, y \textit{ in } I. \tag{4.5.6}$$

(b) If t^{-1} is completely monotonic, then T is TP_2.

Proof. (a) With $u = t(x)$ and $v = t(x) + 2t(y)$, (4.5.6) is equivalent to (4.5.5).

(b) In [Kimberling (1974)] it is proved, by way of Hölder's inequality, that if t^{-1} is c.m. then for all $p > 1$ and all u, v in $\mathbb{R}^+$,

$$[t^{-1}(u+v)]^p \leq t^{-1}(u+pv)[t^{-1}(u)]^{p-1}.$$

With $p = 2$, $u = t(x)$ and $v = t(y)$, this inequality becomes (4.5.6), whence by part (a) and Theorem 4.5.3 T is TP_2. □

The copulas of Example 4.4.8 weakly dominate P, hence are TP_2. The converse of Corollary 4.5.4(b) is therefore false. Note that in view of part (a) of Lemma 4.5.2, part (b) of Corollary 4.5.4 strengthens the last statement in Theorem 4.4.6.

The condition TP_2 admits a natural stochastic interpretation for bivariate distributions and an extension to multivariate distributions as well; see [Nelsen (1999)] and the references cited therein. In this context Theorem 4.5.3 allows the replacement of TP_2 by the ostensibly weaker condition SP_2.

The conditions TP_k and SP_k for $k \geq 3$ are more complicated and, for t-norms, largely unexplored. We state without proof the characterization of SP_3 continuous Archimedean t-norms T given in [Alsina and Tomás (1992)]. Note that since SP_3 implies SP_2, such T must be TP_2 strict copulas. Now given an SP_2 T with generator t, define the function M on $(0,1]^2$ by

$$M(x,y) = \frac{1}{\sqrt{xy}} t^{-1}\left(\frac{t(x)+t(y)}{2}\right).$$

It follows that Ran $M \subseteq (0,1]$ and that (4.5.2), with $m = 3$ and $H = T$, is equivalent to the inequality

$$[1 - M(x,y)^2][1 - M(y,z)^2] \geq [M(x,y)M(y,z) - M(x,z)]^2$$

for all x, y, z in (0,1], which holds if and only if the function $d = \arccos \circ M$ satisfies the triangle inequality. This yields:

Theorem 4.5.5 *A strict copula C with generator t is SP_3 if and only if the function $d : (0,1]^2 \to [0, \pi/2]$ defined by*

$$d(x,y) = \arccos\left[\frac{1}{\sqrt{xy}} t^{-1}\left(\frac{t(x)+t(y)}{2}\right)\right]$$

is a pseudo-metric, in which case

$$C(x,y) = \delta_C(\sqrt{xy}\cos d(x,y)). \tag{4.5.7}$$

When $C = P$, $d \equiv 0$; and when $C = P/(\Sigma - P)$, $d(x,y) = \arccos\left(\frac{2\sqrt{xy}}{x+y}\right)$. Both are pseudo-metrics. There are pseudo-metrics d which cannot arise from an SP_3 strict copula. For instance, substituting $d(x,y) = \frac{\pi}{2}|x-y|$, $x = 1$, and $y = \frac{1}{2}$ in (4.5.7), we obtain $\delta_C\left(\frac{1}{2}\right) = \frac{1}{2}$, which is impossible since C is strict.

Appendix A

Examples and counterexamples

In this section we collect examples which demonstrate unexpected, unusual, or exemplary behavior. We omit verification of the stated properties when they involve straightforward calculations. References to the appropriate sections of the text are given in brackets.

A.1. Independence of the t-norm axioms [§§1.3, 2.1, 2.8]

A.1.1. A continuous binary operation on I that satisfies all of the t-norm conditions except associativity:

$$T(x,y) = xy + xy(1-x)(1-y).$$

A.1.2. Binary operations on I that satisfy all of the t-norm conditions except commutativity:

(a) $$T(x,y) = \begin{cases} x+y-1, & x+y > \frac{3}{2}, \\ \frac{1}{2}, & x+y = \frac{3}{2} \text{ and } x < y, \\ Z(x,y), & \text{otherwise.} \end{cases}$$

(b) $$T(x,y) = \begin{cases} 0, & (x,y) \text{ in } [0,\frac{1}{2}] \times [0,1), \\ \mathrm{Min}(x,y), & \text{otherwise.} \end{cases}$$

A.1.3. A binary operation on I that satisfies all of the t-norm conditions except monotonicity:

$$T(x,y) = \begin{cases} \mathrm{Min}(x,y), & (x,y) \text{ in } \mathbb{Q}^2 \cup (I \backslash \mathbb{Q})^2, \\ 0, & \text{otherwise.} \end{cases}$$

Note that T is nowhere monotonic on $(0,1)^2$, but that $T(x,x) = x$ for all x in I.

A.1.4. Continuous binary operations on I that satisfy all of the t-norm conditions except the boundary conditions:

(a) $T(x,y) = \frac{1}{2}xy$. (b) $T(x,y) = \frac{1}{2}$. (c) $T(x,y) = x + y - 2xy$.
(d) For $0 \le \alpha \le 1$,

$$T_\alpha(x,y) = \begin{cases} \text{Max}(x,y), & (x,y) \text{ in } [0,\alpha]^2, \\ \text{Min}(x,y), & (x,y) \text{ in } [\alpha,1]^2, \\ \alpha, & \text{otherwise.} \end{cases}$$

A.1.5. Continuous binary operations on I that are associative and non-decreasing but not commutative:
(a) $T(x,y) = x$.
(b) For $0 < \alpha < \frac{1}{2}$,

$$T_\alpha(x,y) = \begin{cases} \text{Min}[\text{Max}(x,y),\alpha], & 0 \le x \le \alpha, \\ x, & \alpha \le x \le 1-\alpha, \\ \text{Max}[\text{Min}(x,y),1-\alpha], & 1-\alpha \le x \le 1. \end{cases}$$

Note that $\text{Min} < T_\alpha < \text{Max}$ and $T_\alpha^* = T_\alpha$.

A.1.6. Binary operations on I that satisfy all of the t-norm conditions except that the identity is $\frac{1}{2}$ rather than 1:
(a) Discontinuities only on the boundary: Figure 2.7.1.
(b) Interior discontinuities:

$$T(x,y) = \begin{cases} \text{Max}(x,y), & (x,y) \text{ in } \left[\frac{1}{2},1\right]^2, \\ \text{Min}(x,y), & \text{otherwise.} \end{cases}$$

A.1.7. Continuous, associative, commutative binary operations on I that satisfy exactly one of the t-norm boundary conditions but are not monotonic: Example 2.1.3.

A.1.8. A continuous, associative, commutative, non-decreasing binary operation on [0,1) or (0,1] with interior identity 1/2:

$$T(x,y) = xy/(1 - x - y + 2xy).$$

A.2. Properties of copulas [§§1.4, 2.2]

A.2.1. Families of commutative, non-associative copulas:
(a) $C_\alpha(x,y) = xy[1 + \alpha(1-x)(1-y)]$, $-1 \le \alpha \le 1$ $(a \ne 0)$.
(b) $C_\alpha(x,y) = \frac{1}{1+\alpha}\,\text{Max}[x + y - 1 + \alpha - \alpha|x-y|, 0]$, $0 < \alpha < 1$.
(c) $C_\alpha = (1-\alpha)W + \alpha\text{Min}$, $0 < \alpha < 1$.
(d) $C_\alpha = \frac{\alpha^2(1-\alpha)}{2}W + (1-\alpha^2)P + \frac{\alpha^2(1+\alpha)}{2}\text{Min}$, $-1 < \alpha < 1$ $(\alpha \ne 0)$.

A.2.2. Continuous, Archimedean t-norms that are not copulas but are stronger than W:

(a) Strict: generate T by $t(x) = \begin{cases} \frac{3}{2x} - 1, & 0 \le x \le \frac{1}{2}, \\ \frac{2}{x} - 2, & \frac{1}{2} \le x \le 1. \end{cases}$

(b) Non-strict: generate T by $t(x) = \begin{cases} 5 - 8x, & 0 \le x \le \frac{1}{4}, \\ \frac{1}{x} - 1, & \frac{1}{4} \le x \le 1. \end{cases}$

(c) Non-strict: $T(x,y) = \max\left(\frac{x+y-1}{x+y-xy}, 0\right)$, generated by

$$t(x) = \arctan \sqrt{3}\frac{1-x}{1+x}.$$

(See Corollary 2.2.4 and Theorem 2.2.9.)

A.2.3. A one-parameter family of Archimedean t-norms which are copulas whose limit is a non-trivial ordinal sum:
For $1 < \alpha < \infty$, generate T_α by

$$t_\alpha(x) = \begin{cases} 1 + \frac{\alpha}{2} - \alpha x, & 0 \le x \le \alpha/2(\alpha - 1), \\ 1 - x, & \alpha/2(\alpha - 1) \le x \le 1. \end{cases}$$

T_∞ is the ordinal sum of $\left([0, \frac{1}{2}], W\right)$ and $\left([\frac{1}{2}, 1], W\right)$.

A.3. Properties of discontinuous t-norms [§2.5]

A.3.1. A non-Archimedean t-norm T for which $T(x,x) < x$ for all x in (0,1):

$$T(x,y) = \begin{cases} 0, & (x,y) \text{ in } \left[0, \frac{1}{2}\right]^2, \\ xy + (1-x)(1-y), & (x,y) \text{ in } (\frac{1}{2}, 1]^2, \\ \mathrm{Min}(x,y), & \text{otherwise.} \end{cases}$$

A.3.2. An Archimedean t-norm T for which $\delta_T\left(\frac{1}{2}+\right) = \frac{1}{2}$:

$$T(x,y) = \begin{cases} 0, & (x,y) \text{ in } \left[0, \frac{1}{2}\right]^2, \\ \frac{1}{2}, & (x,y) \text{ in } (\frac{1}{2}, 1)^2, \\ \mathrm{Min}(x,y), & \text{otherwise.} \end{cases}$$

A.3.3. A t-norm that is strictly increasing on $(0,1]^2$ but is not Archimedean: Example 2.5.2.

A.3.4. Discontinuous, strictly increasing, Archimedean t-norms:
(a) Discontinuity only on the boundary: Example 2.5.4(a).

(b) Interior discontinuities:

$$T(x,y) = \begin{cases} \mathrm{Min}(x,y), \; x = 1 \text{ or } y = 1, \\ \frac{1}{4}xy, & (x,y) \text{ in } \left[0, \frac{1}{2}\right]^2, \\ \frac{1}{2}xy, & (x,y) \text{ in } (\frac{1}{2}, 1]^2, \\ \frac{1}{2\sqrt{2}}xy, & \text{otherwise.} \end{cases}$$

A.3.5. A discontinuous Archimedean t-norm whose diagonal is continuous: Example 2.5.7.

A.3.6. A non-measurable t-norm:
Let N be a non- (Borel or Lebesgue) measurable subset of $\left(\frac{1}{2}, 1\right)$ that is symmetric about 3/4. Define T by

$$T(x,y) = \begin{cases} 1/2, & x + y > 3/2 \text{ and } x, y \text{ in } (0,1), \\ 1/2, & x + y = 3/2 \text{ and } x \text{ not in } N, \\ Z(x,y), & \text{otherwise.} \end{cases}$$

(See [Klement, 1982].)

A.3.7. Discontinuous t-norms which coincide with continuous t-norms except in one region where they are constant: Examples 2.6.3(a) and 2.6.3(c).

A.3.8. A discontinuous t-norm T_α, $0 < \alpha < 1$, such that $T_\alpha\left([0,1)^2\right) = \{\alpha, 0\}$: Example 2.6.3(b).

A.3.9. A discontinuous t-norm obtained by truncation: Example 2.6.3(e).

A.3.10. A t-norm T which is continuous except at $(\frac{1}{2}, \frac{1}{2})$, has idempotent $\frac{1}{2}$, and is strictly increasing whenever $T(x,y) \neq 0$ or $\frac{1}{2}$:

$$T(x,y) = W(x,y)/[1 - 4(1-x)(1-y)] \quad \text{for } (x,y) \neq \left(\frac{1}{2}, \frac{1}{2}\right),$$

and $T\left(\frac{1}{2}, \frac{1}{2}\right) = \frac{1}{2}$.

A.3.11. A family of strict t-norms whose limit is a strictly increasing, discontinuous t-norm: Family (2.6.16) as $\alpha \to 0$.

A.4. Representations [§2.7]

A.4.1. A family of discontinuous t-norms which admit continuous outer additive generators:

Example 2.5.4(b), where for $1 < \lambda < 2$,

$$t_{\lambda}^{(-1)}(u) = \begin{cases} 1, & 0 \le u \le \lambda - 1, \\ \lambda - u, & \lambda - 1 \le u \le \lambda, \\ 0, & \lambda \le u \le \infty. \end{cases}$$

A.4.2. A family of discontinuous associative functions which admit continuous inner additive generators:
For $1 < \alpha < 2$,

$$B_{\alpha}(x,y) = \begin{cases} 0, & x + y < \alpha, \\ W(x,y), & x + y \ge \alpha, \end{cases} = b_{\alpha}^{(*)}\left(b_{\alpha}(x) + b_{\alpha}(y)\right),$$

where

$$b_{\alpha}(x) = \begin{cases} 2 - \alpha, & 0 \le x \le \alpha - 1, \\ 1 - x, & \alpha - 1 \le x \le 1, \end{cases}$$

and

$$b_{\alpha}^{(*)}(u) = \begin{cases} 1 - u, & 0 \le u \le 2 - \alpha, \\ 0, & 2 - \alpha < u \le \infty. \end{cases}$$

A.4.3. A continuous, non-decreasing function $F : I^2 \to I$ of the form $F(x,y) = f^{(-1)}(f(x) + f(y))$, which is not associative:

$$F(x,y) = \min(2xy, 1) = f^{(-1)}(f(x) + f(y)),$$

where

$$f(x) = -\log 2x, \quad 0 \le x \le 1,$$

and

$$f^{(-1)}(u) = \begin{cases} 1, & u \le -\log 2, \\ \frac{1}{2}e^{-u}, & u \ge -\log 2. \end{cases}$$

(Note that F has null element 0 and identity $\frac{1}{2}$.)

A.5. Simultaneous associativity [§3.1]

A.5.1. A binary operation T on I that satisfies all of the t-norm conditions except monotonicity and for which $T(x,y) + T^*(x,y) = x + y$:

$$T(x,y) = \sum_{n=1}^{\infty} \frac{x_n y_n}{2^n},$$

where $x = \sum_{n=1}^{\infty} x_n/2^n$, $y = \sum_{n=1}^{\infty} y_n/2^n$ are well-defined dyadic expansions of x, y with $x_i, y_i = 0$ or 1. (Note that $\delta_T = j_I$.)

A.5.2. A family of associative copulas $\{C_\theta\}$ having the Kimeldorf-Sampson properties (i)-(v):
For θ in $[0, \infty]$, generate C_θ via

$$t_\theta(x) = (1 - x)\left(\frac{\log x}{x - 1}\right)^\theta .$$

(Whether C_θ is elementary is an open question.)

A.5.3. Non-associative copulas C for which $C(x, y) = y - C(1 - x, y)$:
(a) Non-commutative: $C(x, y) = \text{Max}[\text{Min}(x, y/2), x + y - 1]$.
(b) Commutative: $C = \sum_{i=1}^{n} \lambda_i (C_{\alpha_i} + C_{-\alpha_i})$, where for all i, α_i belongs to $\mathbb{R}\backslash\{0\}$, $\lambda_i > 0$ with $\sum_{i=1}^{n} \lambda_i = 1/2$, and C_{α_i} is a member of the family (3.1.3). In the case $\lambda_1 = 1/2$, $\alpha_1 = +\infty$ we have the example $(1/2)(W + M)$.

A.6. Duality [§3.2]

A.6.1. A t-norm and an s-norm which cannot be n-duals:
(a) Strict: Example 3.2.5.
(b) Non-strict: $T = W$, $S = \text{Min}\left(\sqrt{x^2 + y^2}, 1\right)$.

A.6.2. A t-norm and an s-norm which are n-dual by way of a unique involution: Example 3.2.6.

A.6.3. A t-norm and an s-norm which are n-dual by way of an infinite family of involutions:
(a) Uncountable family: Example 3.2.7.
(b) Countable family: Example 3.2.8.

A.6.4. Binary operations T and S on I that satisfy all of the t- and s-norm conditions except monotonicity, such that (i) (I, T, S) is a distributive lattice, (ii) $T(x, y) + S(x, y) = x + y$, but (iii) T and

S are not n-dual for any strong negation n:

$$T(x,y) = \sum_{n=1}^{\infty} \mathrm{Min}(x_n, y_n)/2^n,$$

$$S(x,y) = \sum_{n=1}^{\infty} \mathrm{Max}(x_n, y_n)/2^n,$$

where x and y are as given in A.5.1.

A.7. Sets of uniqueness [§3.8]

A.7.1. A continuous Archimedean t-norm $T \neq W$ for which $\mathcal{Z}(T) = \mathcal{Z}(W)$ and $\delta_T = \delta_W$ on $[0, 2/3]$:
Generate T by

$$t(x) = \begin{cases} -3x^2 + 1, & 0 \leq x \leq 1/3, \\ 6x^2 - 6x + 2, & 1/3 \leq x \leq 1/2, \\ -6x^2 + 6x - 1, & 1/2 \leq x \leq 2/3, \\ 3x^2 - 6x + 3, & 2/3 \leq x \leq 1. \end{cases}$$

A.7.2. A continuous Archimedean t-norm $T \neq W$ for which $\mathcal{Z}(T) = \mathcal{Z}(W)$ and $\delta_T = \delta_W$ on $\left[\frac{5}{8}, 1\right]$: Example 3.8.14.

A.7.3. Distinct strict t-norms which have denumerably many common level sets: Example 3.8.15.

A.7.4. Distinct strict t-norms which have denumerably many common sections: See Corollary 3.8.18.

A.7.5. A diagonal and section that cannot be extended to a t-norm:

$$\delta(x) = x^3, \quad \frac{1}{2} \leq x \leq 1; \quad \sigma_{1/2}(x) = \frac{1}{4}(3x - 1), \quad \frac{1}{3} \leq x \leq 1.$$

A.7.6. Distinct strict copulas having common diagonals: The family $\{C_\beta\}$ presented in the discussion following Theorem 3.8.2.

A.8. Concavity and convexity [§4.1]

A.8.1. A non-strict Archimedean t-norm which is Schur-concave but not quasi-concave: Example 4.1.8.

A.8.2. A non-strict Archimedean t-norm which is Schur-concave but not a copula: Example 4.1.9.

A.8.3. A quasi-convex, non-strict Archimedean t-norm whose generator is not concave: Example 4.1.16.

A.8.4. A non-strict Archimedean t-norm which is Schur-convex but not quasi-convex:
Generate T by

$$t(x) = \begin{cases} 1-x, & 0 \le x \le \frac{1}{3}, \\ \frac{1}{2}(1-x)(1+3x), & \frac{1}{3} \le x \le 1. \end{cases}$$

The level curve $T(x,y) = \frac{1}{4}$ is not concave.

A.8.5. A non-strict Archimedean t-norm which is quasi-convex but not convex: T_{-2} from Family (2.6.1).

A.8.6. A commutative copula which is not quasi-concave: $\frac{1}{2}(W + \mathrm{Min})$.

A.8.7. A quasi-concave, non-associative copula: $\frac{1}{2}(P + \mathrm{Min})$.

A.8.8. A continous Archimedean t-norm $T > W$ which is superadditive but not a copula:
Generate T by

$$t(x) = \begin{cases} 1-x, & 0 \le x \le 1/2, \\ 2(1-x)^2, & \frac{1}{2} \le x \le 1. \end{cases}$$

A.8.9. A strict copula C which is not superadditive:
Generate C by

$$t(x) = \begin{cases} e^{-4}\left(1 - \frac{4}{3}\log 4x\right), & 0 \le x \le \frac{1}{4}, \\ e^{-3/(1-x)}, & \frac{1}{4} \le x \le \frac{1}{2}, \\ 2e^{-6}(1-x), & \frac{1}{2} \le x \le 1. \end{cases}$$

Then

$$C\left(\frac{3}{8}, \frac{3}{8}\right) + C\left(\frac{3}{8}, \frac{3}{8}\right) = \frac{18 - 10\log 2}{24 - 5\log 2} > \frac{1}{2} = C\left(\frac{3}{4}, \frac{3}{4}\right).$$

A.9. Order relations and operations [§§1.3, 2.7, 4.2]

A.9.1. Archimedean t-norms T_1, T_2 for which

$$\delta_{T_1} < \delta_W < \delta_{T_2} \quad \text{and } \mathcal{Z}(T_1) \subset \mathcal{Z}(W) \subset \mathcal{Z}(T_2):$$

Generate T_1, T_2 by

$$t_1(x) = \begin{cases} 1 - \sqrt{\frac{x}{2}}, & 0 \le x \le \frac{1}{2}, \\ 1 - x, & \frac{1}{2} \le x \le 1, \end{cases}$$

$$t_2(x) = \begin{cases} 1 - 2x^2, & 0 \le x \le \frac{1}{2}, \\ 1 - x, & \frac{1}{2} \le x \le 1. \end{cases}$$

A.9.2. Continuous t-norms T_1, T_2 for which $\text{Max}(T_1, T_2)$, $\frac{1}{2}(T_1 + T_2)$, and $\text{Min}(T_1, T_2)$ are not associative:
(a) Strict: Any two members of Family (2.6.10).
(b) Non-Archimedean: Any two ordinal sums $([0, a], T_3)$, $([b, 1], T_4)$, where T_3, T_4 are continuous and $0 < b < a < 1$.

A.9.3. A t-norm T for which $\text{Max}(T, W)$ is not a t-norm:

$$T(x, y) = \begin{cases} 3/4, & \text{if } 3/4 \le x, y < 1, \\ Z(x, y), & \text{otherwise.} \end{cases}$$

A.9.4. A t-norm which is a convex combination of non-associative copulas: For the family in Example A.2.1(a), $P = C_0 = \frac{\beta}{\alpha+\beta} C_{-\alpha} + \frac{\alpha}{\alpha+\beta} C_\beta$, for α, β in (0,1).

A.9.5. Elements of $\mathcal{G}_{St}[1, \infty]$ which are convex combinations of elements of $\mathcal{G}_{St}[1, \infty]$:
For $0 \le \alpha \le 1$, $G_\alpha(x, y) = \alpha(x + y - 1) + (1 - \alpha)xy$, generated by $g_\alpha(x) = \log(\alpha + x - \alpha x)$.
(Compare to Family (2.6.7).)

A.9.6. Strict t-norms whose geometric mean is a t-norm:
For Family (2.6.9), $(T_\alpha T_\beta)^{1/2} = T_{\frac{1}{2}(\alpha+\beta)}$.

A.9.7. Strict t-norms whose harmonic mean is a t-norm:
For Family (2.6.3), $1/T_\alpha + 1/T_\beta = 2/T_{\frac{1}{2}(\alpha+\beta)}$.

A.9.8. Commutative and associative binary operations for which dominance is not transitive: Example 4.2.3.

A.9.9. Strict t-norms T_1, T_2 for which T_1 weakly dominates, but does not dominate, T_2: Example 4.2.6.

A.10. Serial iterates and n-copulas [§4.4]

A.10.1. A strict copula $C > P$ whose serial iterates C^n, $n \ge 3$, are not copulas: Example 4.4.8(b).

A.11. Positivity [§4.5]

A.11.1. Associative copulas which are totally positive of order 2 but whose outer additive generators are not completely monotonic: Example 4.4.8.

Appendix B

Open problems

In this section we collect some of the most important unsolved problems concerning t-norms, most of which have been discussed in the text. A subcollection appears in the authors' survey article [Alsina, Frank and Schweizer (2003)]. For additional problems , see [Mesiar and Novák (1996); Klement, Mesiar and Pap (2004d)]. References to the text are given in brackets.

1. [§2.1.] Is part (b) of Lemma 2.1.1 true when assumption (iv) is weakened to continuity in each place?
2. [§2.2] If a t-norm T satisfies the Lipschitz condition (2.2.14) with $\lambda > 1$, then:
   (a) What can one say about T?
   (b) What is the appropriate extension of Theorem 2.2.10 under less restrictive assumptions on t?
3. [§2.5] Characterize the class of t-norms having continuous diagonals.
4. [§2.8] Is the arithmetic mean, or any convex combination, of two distinct t-norms ever a t-norm? More specifically, can P be expressed as a convex combination of associative copulas?
5. [§3.1] Find a simpler, elementary proof of Theorem 3.1.1(a) when T is strict.
6. [§3.1] Find other elementary families of Archimedean copulas that satisfy the Kimeldorf-Sampson properties (i)-(v).
7. [§3.2] Find all t-norms T, s-norms S and strict involutions n such that:
   (a) $S(T(x,y), T(x,n(y))) \leq x$. (Compare to Theorem 3.2.13.)

(b) The function A defined by

$$A(x, y) = S(T(x, n(y)), T(n(x), y))$$

is associative. Note that A corresponds to the symmetric difference.

8. [§3.2] Find all t-norms T_1, T_2 and T_3 such that $T_1 < T_2 < T_3$ and

$$T_3(x, y) - T_2(x, y) = T_2(x, 1 - y) - T_1(x, 1 - y).$$

Note that the special case $T_2 = P$ is solved in Theorem 3.1.5. See also the discussion following the proof of Theorem 3.2.13.

9. [§3.2] If T is a continuous t-norm and the function d_T defined in (3.2.24) is a metric, is T necessarily a copula?
10. [§3.8] The discussion subsequent to Theorem 3.8.2 regarding the copula of two random variables suggest the following questions:
(a) Are there any statistical properties of two random variables which assure that their copula is Archimedean or, more generally, associative?
(b) Can one design a test of statistical independence based on the assumptions that the copula in question is Archimedean and that $F_M(x) = x^2$?
11. [§3.8] Characterize those diagonals of Archimedean t-norms that can be extended to copulas.
12. [§3.8] Theorems 3.8.20, 21, and 23 are particular instances of the following: Let f and g be distinct continuous functions from I into I. If T_1 and T_2 are continuous Archimedean t-norms such that $T_1(x, f(x)) = T_2(x, f(x))$ and $T_1(x, g(x)) = T_2(x, g(x))$ for all x, then $T_1 = T_2$. In other words, the union of the graphs of f and g are sets of uniqueness for $\mathcal{T}_{Ar}$. Find other, or more general, conditions on f and g. In particular, what if $f = j_I$ and $g(x) = ax$ for some a in (0,1)?
13. [§3.8] Find necessary and sufficient conditions for the compatibility of restrictions of the diagonal and a section, i.e., solve the problem of extension for continuous Archimedean t-norms. (See the discussion following Theorem 3.8.21.)
14. [§4.1] Is every strict superadditive t-norm a copula?
15. [§4.1] Characterize continuous, convex t-norms by way of their additive generators.
16. [§4.2] Is the dominance relation (4.2.1) transitive, and hence a partial order, on the set $\mathcal{T}$ of all t-norms? If not, for what subsets is

this the case? This leads to a subsidiary question concerning Mulholland's condition (i): Are there any conditions on the generators t_1 and t_2 of strict t-norms which imply that $\log \circ t_1 \circ t_2^{-1} \circ \exp$ is convex? Also, is the weak dominance relation (4.2.9) transitive on $\mathcal{T}$?

17. [§4.3] Is the associativity equation stable in the sense of Hyers?
18. [§4.4] Explore the connection between Archimedean, non-strict n-copulas and the n-monotonicity of their generators $t^{(-1)}$ in general, and for one-parameter families in particular.
19. [§4.5] Among strict copulas, is SP_3 a stronger condition than TP_2?
20. The space of all two-dimensional copulas, endowed with the L_∞-metric, is complete. Is the set of associative copulas a set of first category in this space?

Bibliography

Abel, N.H. (1826). Untersuchung der Functionen zweier unabhängig veränderlichen Grössen x und y wie $f(x,y)$, welche die Eigenschaft haben, dass $f(z,f(x,y))$ eine symmetrische Function von x, y und z ist. *J. Reine Angew. Math.* **1**, pp. 11-15. *Oeuvres Complètes de N.H. Abel, Vol. 1.* Christiana, 1881, pp. 61-65.

Aczél, J. (1949). Sur les opérations définies pour nombres réels. *Bull. Soc. Math. France* **76**, pp. 59-64.

Aczél, J. (1966). *Lectures on Functional Equations and Their Applications.* Academic Press, New York.

Aczél, J. (1987). *A Short Course on Functional Equations.* Reidel, Dordrecht.

Aczél, J. (1989). The state of the second part of Hilbert's fifth problem. *Bull. Amer. Math. Soc.* **20**, pp. 153-163.

Aczél, J. and Alsina, C. (1984). Characterizations of some classes of quasilinear functions with applications to triangular norms and to synthesizing judgements. *Methods of Operations Research* **48**, pp. 3-22.

Aczél, J. and Alsina, C. (1986). On synthesis of judgements. *Socio-Econ. Plan. Sci.* **20**, 6, pp. 333-339.

Aczél, J. and Alsina, C. (1987). Synthesizing judgements: a functional equation approach. *Math. Modelling* **9**, pp. 311-320.

Aczél, J. and Dhombres, J. (1989). *Functional Equations in Several Variables.* Cambridge Univ. Press, Cambridge.

Aczél, J. and Saaty, T.L. (1983). Procedures for synthesizing ratio judgements. *J. Math. Psych.* **27**, pp. 93-102.

Agell, N. (1984). Sobre la concavidad de t-normas y de funciones triangulares. *Stochastica* **VIII**, pp. 91-95.

Alsina, C. (1978). On countable products and algebraic convexifications of probabilistic metric spaces. *Pacific J. Math.* **76**, pp. 291-300.

Alsina, C. (1980). On a family of functional inequalities. In E.F.Beckenbach (Ed.). *General Inequalities* **2**. Birkhäuser Verlag, Basel, pp. 419-427.

Alsina, C. (1981). Some functional equations in the space of uniform distribution functions. *Aequationes Math.* **22**, pp. 153-164.

Alsina, C. (1984a). On Schur-concave t-norms and triangle functions. In

W. Walter (Ed.), *General Inequalities* **4**. Birkhäuser Verlag, Basel, pp. 241-248.

Alsina, C. (1984b). On some metrics induced by copulas. *Ibid*, p. 397.

Alsina, C. (1985a). Characterization of some topological semigroups whose level curves are homothetic. *Rend. Circ. Mat. di Palermo Serie II*, **XXXIV**, pp. 136-140.

Alsina, C. (1985b). On a family of connectives for fuzzy sets. *Fuzzy Sets and Systems* **16**, pp. 231-235.

Alsina, C. (1986a). On associative developable surfaces. *Archiv. Math. (Brno)* **22**, pp. 93-96.

Alsina, C. (1986b). The associative solutions of the functional equation $\tau(F,G) + \hat{\tau}(F,G) = F + G$. *Utilitas Math.* **29**, pp. 93-98.

Alsina, C. (1988). On a functional equation characterizing two binary operations on the space of membership functions. *Fuzzy Sets and Systems* **27**, pp. 5-9.

Alsina, C. (1991). On the stability of a functional equation related to associativity. *Ann. Polon. Math.* **53**, pp. 1-5.

Alsina, C. (1992). On a method of Pi-Calleja for describing additive generators of associative functions. *Aequationes Math.* **43**, pp. 14-20.

Alsina, C. (1996). As you like them: connectives in Fuzzy Logic. *Proc. Int. Symp. Multi-Valued Logic '96*, Univ. Santiago de Compostela, pp. 1-7.

Alsina, C. (1997). On connectives in fuzzy logic satisfying the condition $S(T_1(x,y), T_2(x, N(y))) = x$. *Proc. FUZZ-IEEE 1997*, Barcelona, pp. 149-153.

Alsina, C., Frank, M.J. and Schweizer, B. (2003). Problems on associative functions. *Aequationes Math.* **66**, pp. 128-140.

Alsina, C. and Ger, R. (1985). Associative operations close to a given one. *Comptes Rendus Math. Rep. Acad. Sci. Canada* **VII**, pp. 207-210.

Alsina, C. and Ger, R. (1988). On associative copulas uniformly close. *Int. J. Math. Math. Sci.* **II**, pp. 439-448.

Alsina, C. and Giménez, J. (1984). Sobre L-órdenes entre t-normas estrictas. *Stochastica* **VIII**, 1, pp. 85-89.

Alsina, C., Nelsen, R.B. and Schweizer, B. (1993). On the characterization of a class of binary operations on distribution functions. *Statist. Probab. Letters* **17**, pp. 85-89.

Alsina, C. and Sklar, A. (1987). A characterization of continuous associative operations whose graphs are ruled surfaces. *Aequationes Math.* **33**, pp. 114-119.

Alsina, C. and Tomás, M.S. (1988). Smooth convex t-norms do not exist. *Proc. Amer. Math. Soc.* **102**, pp. 317-320.

Alsina, C. and Tomás, M.S. (1992). On positive semidefinite strict t-norms. *Int. Ser. Num. Math.* **103**, pp. 214-225.

Alsina, C. and Trillas, E. (1984). Sobre operadors d'implicació en lògica polivalent. *Actes III Congrès Català de Lògica Matemàtica*, Pub. Univ. Politècnica de Catalunya, pp. 59-61.

Alsina, C. and Trillas, E. (1985). On some indistinguishability operators. *Proc. XVth. Int. Symp. Multiple-Valued Logic*, Kingston (Canada), pp. 154-155.

Alsina, C. and Trillas, E. (1987). Additive homogeneity of logical connectives for membership functions. In Bezdek, J.C. (Ed.) *Analysis of Fuzzy Information* **I**. CRC Pubs., Boca Raton, pp. 179-183.

Alsina, C. and Trillas, E. (1992). On almost distributive Lukasiewicz triplets. *Fuzzy Sets and Systems* **50**, pp. 175-178.

Alsina, C., Trillas, E. and Valverde, L. (1980). On non-distributive logical connectives for fuzzy set theory. *BUSEFAL* **3**, pp. 18-29.

Alsina, C., Trillas, E. and Valverde, L. (1982). Do we need Max, Min and 1-j in fuzzy set theory? In R.R. Yager (Ed.), *Fuzzy Set and Possibility Theory: Recent Developments*. Pergamon, New York, pp. 275-297.

Alsina, C., Trillas, E. and Valverde, L. (1983). On some logical connectives for fuzzy sets theory. *J. Math. Anal. Appl.* **93**, pp. 15-26.

Alt, W. (1940). Über die reellen Funktionen einer reellen Veränderlichen, welche ein rationales Additionstheorem besitzen. *Deutsche Math.* **5**, pp. 1-12.

Apostol, T. (1974). *Mathematical Analysis, 2nd ed.* Addison-Wesley, Reading MA.

Arnold, V.I. (1957). Concerning the representability of functions of two variables in the form $\chi[\phi(x) + \psi(y)]$. *Uspehi Mat. Nauk* **12**, (74), pp. 119-121.

Bacchelli, B. (1986). Representation of continuous associative functions. *Stochastica* **X**, pp. 13-28.

Bandler, W. and Kohout, L. (1980). Fuzzy power sets and fuzzy implication operators. *Fuzzy Sets and Systems* **4**, pp. 13-30.

Beckenbach, E.F. and Bellman, R. (1965). *Inequalities, 2nd rev. ed.* Springer Verlag, Berlin.

Behringer, F.A. (1980). More on Karamardian's theorem concerning the quasiconvexity of strictly quasiconvex functions. *Z. Angew. Math. Mech.* **60**, pp. 201-202.

Bell, E.T. (1936). A functional equation in arithmetic. *Trans. Amer. Math. Soc.* **39**, pp. 341-344.

Beneš, V. and Štěpán, J. (Eds.) (1997). *Distributions with Given Marginals and Moment Problems*. Kluwer, Dordrecht.

Bertoluzza, C. (1977). Functional equations in the theory of fuzzy sets. *Aequationes Math.* **15**, pp. 273-274.

Bézivin, J.P. and Tomás, M.S. (1993). On the determination of strict t-norms on some diagonal segments. *Aequationes Math.* **45**, pp. 239-245.

Boas, R.P. (1971). Signs of derivatives and analytic behavior. *Amer. Math. Monthly* **78**, pp. 1085-1093.

Boas, R.P. (1972). *A Primer of Real Functions, 2nd. ed.* Carus Math. Monog. 13. Math. Assoc. of Amer., Wiley, New York.

Bôcher, M. (1907). *Introduction to Higher Algebra*. Macmillan, New York.

Bodiou, G. (1964). *Théorie Dialectique des Probabilités*. Gauthier-Villars, Paris.

Bohnenblust, A. (1940). Axiomatic characterization of L_p-spaces. *Duke Math. J.* **5**, pp. 1-12.

Brouwer, L.E.J. (1909). Die Theorie der endlichen kontinuierlichen Gruppen unabhängig von den Axiomen von Lie. *Math. Ann.* **67**, pp. 246-267.

Burgués, C. (1981). Sobre la sección diagonal y la región cero de una t-norma.

Stochastica **V**, pp. 79-87.

Capéraà, P. and Genest, C. (1990). Concepts de dépendance et ordres stochastiques pour des lois bidimensionelles. *Canad. J. Statist.* **18**, pp. 315-326.

Carruth, J.H., Hildebrant, J.A. and Koch, R.J. (1983). *The Theory of Topological Semigroups.* Marcel Dekker, New York.

Cartan, E. (1930). La théorie des groupes finis et continus et l'analyse situs. *Mem. Sci. Math.* **42**, Gauthier-Villars, Paris.

Cherubini, U., Luciano, E. and Vecchiato, W. (2004). *Copula Methods in Finance.* Wiley, Hoboken, NJ.

Chow, Y.S. and Teicher, H. (1997). *Probability Theory: Independence, Interchangeability, Martingales (3rd ed.).* Springer Verlag, New York.

Clifford, A.H. (1958). Totally ordered commutative semigroups. *Bull. Amer. Math. Soc.* **64**, pp. 305-316.

Clifford, A.H. and Preston, G.B. (1961, 1967). *The Algebraic Theory of Semigroups, Vols. I, II.* Amer. Math. Soc., Providence.

Climescu, A.C. (1946). Sur l'équation fonctionnelle de l'associativité. *Bul. Inst. Politehn. Iaşi* **1**, pp. 1-16.

Cooper, R. (1927). Notes on certain inequalities (II). *J. London Math. Soc.* (2) **26**, pp. 415-432.

Craigen, R. and Páles, Z. (1989). The associativity equation revisited. *Aequationes Math.* **37**, pp. 306-312.

Cuadras, C.M., Fortiana, J. and Rodriguez-Lallena, J.A. (Eds.) (2002). *Distributions with Given Marginals and Statistical Modeling.* Kluwer, Boston.

Dall'Aglio, G. (1960). Les fonctions extrême de la classe de Fréchet à 3 dimensions. *Publ. Inst. Statist. Univ. Paris* **9**, pp. 175-188.

Dall'Aglio, G. (1972). Fréchet classes and compatibility of distribution functions. *Symposia Math.* **9**, pp. 131-150.

Dall'Aglio, G. (1991). Fréchet classes: the beginnings. In [Dall'Aglio, et al. (1991)], pp. 1-12.

Dall'Aglio, G., Kotz, S. and Salinetti, G. (Eds.) (1991). *Advances in Probability Distributions with Given Marginals; Beyond the Copulas.* Mathematics and its Applications, v. 67, Kluwer, Dordrecht.

Daróczy, Z. and Losonczi, L. (1967). Über die Erweiterung der auf einer Punktmenge additiven Functionen. *Publ. Math. Debrecen* **14**, pp. 239-245.

Daróczy, Z. and Páles, Z. (1987). Convexity with given infinite weight sequences. *Stochastica* **XI**, pp. 5-12.

Darsow, W.F. and Frank, M.J. (1983). Associative functions and Abel-Schröder systems. *Publ. Math. Debrecen* **30**, pp. 253-272.

De Rham, G. (1956). Sur quelques courbes définies par des equations fonctionnelles. *Rend. Sem. Mat. Univ. Politec. Torino* **16**, pp. 101-113. (Translated and reprinted as "On some curves defined by functional equations" in Edgar, G.A. (Ed.) (1993), *Classics on Fractals.* Addison-Wesley, Reading, MA, pp. 285-297).

Dhombres, J. (1979). *Some Aspects of Functional Equations.* Chulalongkorn Univ. Press, Bangkok.

Dickson, L.E. (1916). An extension of the theory of numbers by means of correspondences between fields. *Bull. Amer. Math. Soc.* **23**, pp. 109-111.

Drouet Mari, D. and Kotz, S. (2001). *Correlation and Dependence.* Imperial Coll. Press, London.

Dubois, D. (1980). Triangular norms for fuzzy sets. In E.P. Klement (Ed.), *Proc. Second Int. Sem. Fuzzy Set Theory.* Linz, pp. 39-68.

Dubois, D. and Prade, H. (1980). *Fuzzy Sets and Systems: Theory and Applications.* Academic Press, New York.

Ebanks, B.R. (1998). Quasi-homogeneous associative functions. *Int. J. Math. Math. Sci.* **21**, pp. 351-358.

Faucett, W.M. (1955). Compact semigroups irreducibly connected between two idempotents. *Proc. Amer. Math. Soc.* **6**, pp. 741-747.

Feller, W. (1966). *An Introduction to Probability Theory and its Applications. Vol. II.* Wiley, New York.

Fenchel, W. (1953). *Convex Cones, Sets and Functions.* Lecture Notes, Princeton Univ., Princeton.

Fenyö, I. and Paganoni, L. (1987). Su una regola di addizione razionale. *Rend. Sem. Mat. Univ. Politec. Torino* **45**, pp. 105-116.

de Finetti, B. (1949). Sulle stratificazioni convesse. *Ann. Mat. Pura Appl.* **(4) 30**, pp. 173-183.

Fodor, J.C. and Roubens, M. (1994a). Valued preference structures. *Eur. J. Oper. Res.* **79**, pp. 277-286.

Fodor, J.C. and Roubens, M. (1994b). *Fuzzy Preference Modelling and Multicriteria Decision Support.* Kluwer, Dordrecht.

Fodor, J.C., Yager, R.R. and Rybalov, A. (1997). Structure of uninorms. *Int. J. Uncertainty, Fuzziness and Knowledge-Based Systems* **5**, pp. 411-427.

Frank, M.J. (1975). Associativity in a class of operations on spaces of distribution functions. *Aequationes Math.* **12**, pp. 121-144.

Frank, M.J. (1979). On the simultaneous associativity of $F(x,y)$ and $x+y-F(x,y)$. *Aequationes Math.* **19**, pp. 194-226.

Frank, M.J. (1981). An equation which links associative functions. *Abstracts Amer. Math. Soc.* **2**, p. 128.

Frank, M.J. (1991). Convolutions for dependent random variables. In [Dall'Aglio, et al. (1991)], pp. 75-93.

Frank, M.J. (1996). Diagonals of copulas and Schröder's equation. *Aequationes Math.* **51**, p. 150.

Frank, M.J. (2002). Multivariate Archimedean copulas. *Aequationes Math.* **64**, p. 175.

Fréchet, M. (1935). Généralizations du théorème des probabilités totales. *Fund. Math.* **25**, pp. 379-387.

Fréchet, M. (1951, 1957, 1958). Sur les tableaux de corrélation dont les marges sont données. *Ann. Univ. Lyon, Sec. A*, **14**, pp. 53-77; **20**, pp. 13-31; **21**, pp. 19-32.

Fuchs, L. (1963). *Partiallly Ordered Algebraic Systems.* Pergamon Press, Oxford.

Gaines, B.R. (1976). Foundations of fuzzy reasoning. *Internat. J. Man-Machine Stud.* **8**, pp. 623-668.

García, P. and Valverde, L. (1989). Isomorphisms between de Morgan triplets.

Fuzzy Sets and Systems **30**, pp. 27-36.

Genest, C. (1987). Frank's family of bivariate distributions. *Biometrika* **74**, pp. 549-555.

Genest, C. et al. (Eds.) (2005). *Proceedings of the Conference on Dependence Modeling: Statistical Theory and its Applications in Finance and Insurance (DeMoSTAFI)*. In *Insurance: Mathematics and Economics*, vol. **37**, no. 1, and *The Canadian Journal of Statistics*, vol. **33**, no. 3.

Genest, C. and MacKay, R.J. (1986a). The joy of copulas: bivariate distributions with uniform marginals. *Amer. Statist.* **40**, pp. 280-283.

Genest, C. and MacKay, R.J. (1986b). Copules archimédiennes et familles de lois bidimensionnelles dont les marges sont donées. *Canad. J. Statist.* **14**, pp. 145-159.

Genest, C., Quesada, J.J., Rodriguez-Lallena, J.A. and Sempi, C. (1999). A characterization of quasi-copulas. *J. Multivariate Anal.* **69**, pp. 193-205.

Genest, C. and Rivest, L.P. (1989). A characterization of Gumbel's family of extreme value distributions. *Statist. Probab. Letters* **8**, pp. 207-211.

Genest, C. and Rivest, L.P. (1993). Statistical inference procedures for bivariate Archimedean copulas. *J. Amer. Stat. Assn.* **88**, pp. 1034-1043.

Ger, R. (1996). Iterative subsemigroups of the unit interval with a t-norm structure. In C. Mira et al. (Eds.), *Proc. European Conference on Interation Theory (ECIT-92)*. World Scientific Publ. Co., Singapore, pp. 126-135.

Gottwald, S. (1993). *Fuzzy Sets and Fuzzy Logic. Foundations of Application - from a Mathematical Point of View.* Vieweg, Braunschweig/Wiesbaden.

Hadzic, O. and Pap, E. (2001). *Fixed Point Theory in Probabilistic Metric Spaces.* Kluwer, Dordrecht.

Hailperin, T. (1965). Best possible inequalities for the probability of a logical function of sets. *Amer. Math. Monthly* **72**, pp. 343-359.

Hardegree, G.M. (1981). Quasi-implication algebras. I. Elementary theory. II. Structure theory. *Algebra Universalis* **12**, pp. 30-47; 48-65.

Hardy, G.H., Littlewood, J.E. and Pólya, G. (1952). *Inequalities, 2nd. ed.* Cambridge Univ. Press, New York-London.

Hille, E and Phillips, R.S. (1957). *Functional Analysis and Semigroups.* Amer. Math. Soc. Colloq. Publ. 31, Amer. Math. Soc., Providence, RI.

Hoeffding, W. (1940). Masstabinvariante Korrelationstheorie. *Schriften des Mathematischen Instituts und des Instituts für Angewandte Mathematik der Universität Berlin* **5**, Heft 3, pp. 181-233. (Translated and reprinted as "Scale invariant correlation theory" in Fisher, N.I. and Sen, P.K. (Eds.) (1994), *The Collected Works of Wassily Hoeffding.* Springer, New York, pp. 57-107.)

Hofmann, K.H. (1994). Semigroups and Hilbert's fifth problem. *Math. Slovaca* **44**, pp. 365-377.

Hofmann, K.H. and Lawson, J.D. (1996). A survey on totally ordered semigroups. In K.H. Hofmann and W.M. Mislove (Eds.), *Semigroup Theory and its Applications.* London Math. Soc. Lecture Note Series, No. 231, Cambridge Univ. Press, Cambridge, pp. 15-39.

Hofmann, K.H. and Mostert, P.S. (1966). *Elements of Compact Semigroups.*

Merrill Books, Columbus, Ohio.

Höhle, U. and Klement, E.P. (Eds.) (1995). *Non-Classical Logics and Their Applications to Fuzzy Subsets - A Handbook of the Mathematical Foundations of Fuzzy Set Theory.* Kluwer, Dordrecht.

Hutchinson, T.P. and Lai, C.D. (1990). *Continuous Bivariate Distributions, Emphasizing Applications.* Rumsby Scientific Publishing, Adelaide.

Hyers, D.H. (1941). On the stability of the linear functional equation. *Proc. Nat. Acad. Sci. USA* **27**, pp. 222-224.

Jarai, A. and Maksa, G. (1985). Remark to the problem No. 11 of Prof. C. Alsina. *Proc. 23rd Int. Symposium on Functional Equations*, Univ. of Waterloo, Canada, pp. 64-65.

Jarczyk, W. and Matkowski, J. (2002). On Mulholland's inequality. *Proc. Amer. Math. Soc.* **130**, pp. 3243-3247.

Joe, H. (1993). Parametric families of multivariate distributions with given margins. *J. Multivariate Anal.* **46**, pp. 262-282.

Joe, H. (1997). *Multivariate Models and Dependence Concepts.* Chapman & Hall, London.

Jouini, M. and Clemen, R. (1996). Copula models for aggregating expert opinions. *Operations Res.* **44,** pp. 444-457.

Kairies, H.-H. (1997). Functional equations for peculiar functions. *Aequationes Math.* **53**, pp. 207-241.

Kampé de Fériet, J. (1974). La théorie généralisé de l'information et la mesure subjective de l'information. *Lectures Notes in Math. 398.* Springer-Verlag, New York, pp. 1-35, .

Kampé de Fériet, J. and Forte, B. (1967). Information et probabilité. *C.R. Acad. Sci. Paris* **265A**, pp. 110-114, pp. 142-146, pp. 350-353.

Kampé de Fériet, J., Forte, B. and Benvenutti, P. (1969). Forme générale de l'opération de composition continue d'une information. *C.R. Acad. Sci. Paris* **269A**, pp. 529-534.

Karlin, S. (1968). *Total Positivity, Vol. I.* Stanford Univ. Press, Stanford, CA.

Kimberling, C.H. (1973). On a class of associative functions. *Publ. Math. Debrecen* **20**, pp. 21-39.

Kimberling, C.H. (1974). A probabilistic interpretation of complete monotonicity. *Aequationes Math.* **10**, pp. 152-164.

Kimeldorf, G. and Sampson, A.R. (1975a). One-parameter families of bivariate distributions with fixed marginals. *Comm. Statist.* **4**, pp. 293-301.

Kimeldorf, G. and Sampson, A.R. (1975b). Uniform representations of bivariate distributions. *Comm. Statist.* **4**, pp. 617-627.

Klement, E.P. (1980). Some remarks on t-norms, fuzzy σ-algebras and fuzzy mesures. In E.P. Klement (Ed.), *Proc. Second Int. Sem. on Fuzzy Set Theory.* Linz, pp. 125-142.

Klement, E.P. (1982). Construction of fuzzy σ-algebras using triangular norms. *J. Math. Anal. Appl.* **85**, pp. 543-565.

Klement, E.P. and Mesiar, R. (Eds.) (2005). *Logical, Algebraic, Analytic and Probabilistic Aspects of Triangular Norms.* Elsevier, Amsterdam.

Klement, E.P., Mesiar, R. and Pap, E. (2000). *Triangular Norms.* Kluwer, Dordrecht.

Klement, E.P., Mesiar, R. and Pap, E. (2004a). Triangular norms I: Basic analytical and algebraic properties. *Fuzzy Sets and Systems* **143**, pp. 5-26.

Klement, E.P., Mesiar, R. and Pap, E. (2004b). Triangular norms II: General constructions and parametrized families. *Fuzzy Sets and Systems* **145**, pp. 411-438.

Klement, E.P., Mesiar, R. and Pap, E. (2004c). Triangular norms III: Continuous t-norms. *Fuzzy Sets and Systems* **145**, pp. 439-454.

Klement, E.P., Mesiar, R. and Pap, E. (2004d). Problems on triangular norms and related operators. *Fuzzy Sets and Systems* **145**, pp. 471-479.

Koch, R.J. (1957). Note on weak cutpoints in clans. *Duke Math. J.* **24**, pp. 611-615.

Kolesárová, A. (1999). A note on Archimedean triangular norms. *BUSEFAL* **80**, pp. 57-60.

Krause, G.M. (1981). *A strengthened form of Ling's theorem on associative functions.* Ph.D. Thesis, Illinois Institute of Technology, Chicago.

Krause, G.M. (1983). Interior idempotents and non-representability of groupoids. *Stochastica* **VII**, pp. 5-10.

Kruse, R.L. and Deely, J.J. (1969). Joint continuity of monotonic functions. *Amer. Math. Monthly* **76**, pp. 74-76.

Kuczma, M. (1963). *On the Schröder Equation.* Rozprawy Math. 34, Panstowowe Wydawnictwo Naukowe, Warszawa.

Kuczma, M. (1968). *Functional Equations in a Single Variable.* Monografie Mat. 46, Polish Scientific Publ., Warszawa.

Kuczma, M. (1978). Functional equations on restricted domains. *Aequationes Math.* **18**, pp. 1-34.

Kuczma, M. (1985). *An Introduction to the Theory of Functional Equations and Inequalities (Cauchy's Equation and Jensen's Inequality).* Univ. Śląski, Państwowe Wydawnictwo Naukowe, Warszawa-Krakow-Katowice.

Kuczma, M., Choczewski, B. and Ger, R. (1990). *Iterative Functional Equations.* Cambridge Univ. Press, Cambridge.

Lawson, J.D. (1996). The earliest semigroup paper?. *Semigroup Forum* **52**, pp. 55-60.

Levinson, N. and Redheffer, R.M. (1970). *Complex Variables.* Holden-Day, San Francisco.

Ling, C.-H. (1965). Representation of associative functions. *Publ. Math. Debrecen* **12**, pp. 189-212.

López de Mantaras, R. (1990). *Approximate Reasoning Models.* Ellis Horwood series on Artificial Intelligence, Chichester.

Mak, K.-T. and Sigmon, K. (1988). Standard threads and distributivity. *Aequationes Math.* **36**, pp. 251-267.

Maksa, G. (2000). The generalized associativity equation revisited. *Prace Matematyczne* **XVII**, pp. 175-180.

Marco, J. and Ruiz-Rivas, J. (1992). On the construction of multivariate distributions with given nonoverlapping multivariate marginals. *Statist. & Probab.*

Letters **15**, pp. 259-265.

Marichal, J.-L. (1998). On the associativity functional equation. *G.E.M.M.E.*, No. 9805.

Marley, A.A.J. (1982). Random utility models with all choice probabilities expressible as 'functions' of the binary choice probabilities. *Math. Soc. Sci.* **3**, pp. 39-56.

Marshall, A.W. and Olkin, I. (1979). *Inequalities: Theory of Majorization and Its Applications.* Academic Press, New York.

Marshall, A.W. and Olkin, I. (1988). Families of multivariate distributions. *J. Amer. Stat. Assn.* **83**, pp. 834-841.

Matkowski, J. (1992). L^p-like paranorms. *Gräzer Math. Ber.* **316**, pp. 103-138.

Mayor, G. (1994). On a family of quasi-arithmetic means. *Aequationes Math.* **48**, pp. 137-142.

Mayor, G. and Torrens, J. (1992). Duality for a class of binary operations on [0,1]. *Fuzzy Sets and Systems* **47**, pp. 77-80.

Mayor, G. and Torrens, J. (1994). De Rham systems and the solution of a class of functional equations. *Aequationes Math.* **47**, pp. 43-49.

Meester, S.G. and MacKay, J. (1994). A parametric model for cluster correlated categorical data. *Biometrics* **50**, pp. 954-963.

Menger, K. (1942). Statistical metrics. *Proc. Nat. Acad. Sci. USA* **28**, pp. 535-537. (Reprinted in [Schweizer, Sklar, Sigmund, et al. (2003)], pp. 433-435.)

Menger, K. (1951a). Probabilistic theories of relations. *Proc. Nat. Acad. Sci. USA* **37**, pp. 178-180. (Reprinted in [Schweizer, Sklar, Sigmund, et al. (2003)], pp. 437-439.)

Menger, K. (1951b). Ensembles flous et fonctions aleatoires. *C.R. Acad. Sci. Paris* **232**, pp. 2001-2003. (Reprinted in [Schweizer, Sklar, Sigmund, et al. 2003)], pp. 445-447.)

Mesiar, R. and Novák, V. (1996). Open problems from the 2nd Int. Conf. on Fuzzy Sets Theory and Its Applications. *Fuzzy Sets and Systems* **81**, pp. 185-190.

Mostert, P.S. and Shields, A.L. (1957). On the structure of semigroups on a compact manifold with boundary. *Ann. of Math.* **65**, pp. 117-143.

Motzkin, T.S. (1936). Sur le produit des espaces métriques. *C.R. Cong. Intern. Math. Oslo, Vol. II,* pp. 137-138.

Moynihan, R. (1978). On τ_T-semigroups of probability distribution functions, II. *Aequationes Math.* **17**, pp. 19-40.

Mulholland, H.P. (1950). On generalizations of Minkowski's inequality in the form of a triangle inequality. *Proc. London Math. Soc.* **(2) 51**, pp. 294-307.

Nelsen, R.B. (1986). Properties of a one-parameter family of bivariate distributions with specified marginals. *Comm. Statist.* **A 15**, pp. 3277-3285.

Nelsen, R.B. (1991). Copulas and association. In [Dall'Aglio, et al. (1991)], pp. 51-74.

Nelsen, R.B. (1995). Copulas, characterization, correlation, and counterexamples. *Math. Mag.* **68**, pp. 193-198.

Nelsen, R.B. (1997). Dependence and order in families of Archimedean copulas. *J. Multivariate Anal.* **60**, pp. 111-122.

Nelsen, R.B. (1999). *An Introduction to Copulas.* Lecture Notes in Statististics 139. Springer, New York. Revised edition to appear in 2006.

Nelsen, R.B. (2005). Copulas and quasi-copulas: An introduction to their properties and applications. In [Klement and Mesiar (2005)], pp. 391-414.

Nelsen, R.B. and Úbeda-Flores, A. (to appear). The lattice-theoretic structure of sets of bivariate copulas and quasi-copulas. *C.R. Acad. Sci. Paris, Serie I, Mathematique.*

Nguyen, H.T. and Walker, E.A. (2000). *A First Course in Fuzzy Logic, 2nd ed.* CRC Press, Boca Raton.

Niven, J. (1956). *Irrational Numbers.* Carus Math. Monog. 11. Math. Assoc. of Amer., Wiley, New York.

Paalman-de Miranda, A.B. (1964). *Topological Semigroups.* Math. Centre Tracts 11, Mathematisch Centrum Amsterdam.

Paganoni, L. and Rusconi, D. (1983). A characterization of some classes of functions F of the form $F(x,y) = g(\alpha f(x) + \beta f(y) + \gamma)$ or $F(x,y) = \phi(h(x) + k(y))$. *Aequationes Math.* **26**, pp. 138-162.

Pavelka, J. (1979). On Fuzzy Logic I, II, III. *Zeitschr. Math. Logik Grundlagen Math.* **25**, pp. 45-54, 119-134, 447-464.

Pi-Calleja, P. (1954). Las ecuaciones funcionales de la teoría de magnitudes. In *Segundo Symposium de Matemática*, Villavicencio, Mendoza. Coni, Buenos Aires, pp. 199-280.

Roberts, A.W. and Varberg, D.E. (1973). *Convex Functions.* Academic Press, New York.

Rüschendorf, L., Schweizer, B. and Taylor, M.D. (Eds.) (1996). *Distributions with Fixed Marginals and Related Topics.* Institute of Mathematical Statistics, Lect. Notes-Monograph Series Vol. 28, Hayward, CA.

Rusconi, D. (1992). Caratterizzazione di classi di funzioni della forma $F(x,y) = \varphi(h(x) + k(y))$. *Stochastica* **XIII**, pp. 115-136.

Ruspini, R. (1982). Recent developments in fuzzy clustering. In R.R. Yager (Ed.), *Fuzzy Set and Possibility Theory: Recent Developments.* Pergamon, New York, pp. 133-147.

Sánchez-Soler, M. (1988). On the functional equation $H[\tau(F,G), \varsigma(F,G)] = H(F,G)$. *Stochastica* **XII**, pp. 131-139.

Sander, W. (1998). Associativity. In *Leaflets in Mathematics, Proc. Numbers, Functions, Equations '98 Intern. Conf.* Janus Pannonius Univ., Pécs, pp. 139-140.

Schweizer, B. (1991). Thirty years of copulas. In [Dall'Aglio, et al. (1991)], pp. 13-50.

Schweizer, B. (2003). Commentary on probabilistic geometry, in [Schweizer, Sklar, Sigmund, et al. (2003)], pp. 409-432.

Schweizer, B. and Sklar, A. (1960). Statistical metric spaces. *Pacific J. Math.* **10**, pp. 313-334.

Schweizer, B. and Sklar, A. (1961). Associative functions and statistical triangle inequalities. *Publ. Math. Debrecen* **8**, pp. 169-186.

Schweizer, B. and Sklar, A. (1963). Associative functions and abstract semigroups. *Publ. Math. Debrecen* **10**, pp. 69-81.

Schweizer, B. and Sklar, A. (1974). Operations on distribution functions not derivable from operations on random variables. *Studia Math.* **52**, pp. 43-52.

Schweizer, B. and Sklar, A. (1983). *Probabilistic Metric Spaces.* Elsevier-North Holland, New York.

Schweizer, B. and Sklar, A. (2005). Reissue of [Schweizer and Sklar (1983)] with supplementary comments and updated bibliography. Dover Publications, New York.

Schweizer, B., Sklar, A., Sigmund, K., et al. (Eds.) (2002, 2003). *Karl Menger Selecta Mathematica, Volumes 1,2.* Springer-Verlag, Vienna and New York.

Schweizer, B. and Wolff, E.F. (1976). Sur une mesure de dépendance pour les variables aléatoires. *C.R. Acad. Sci. Paris* **283A**, pp. 659-661.

Schweizer, B. and Wolff, E.F. (1981). On nonparametric measures of dependence for random variables. *Ann. Statist.* **9**, pp. 879-885.

Sherwood, H. (1984). Characterizing dominates in a family of triangular norms. *Aequationes Math.* **27**, pp. 255-273.

Shyu, Y.-H. (1984). *Absolute continuity in the* τ_T*-operations.* Ph.D. Thesis, Illinois Institute of Technology, Chicago.

Sklar, A. (1959). Fonctions de répartition à n dimensions et leurs marges. *Publ. Inst. Statist. Univ. Paris* **8**, pp. 229-231.

Sklar, A. (1973). Random variables, joint distribution functions and copulas. *Kybernetika* **9**, pp. 449-460.

Sklar, A. (1987). The structure of one-dimensional flows with continuous trajectories. *Radovi Mat.* **3**, pp. 111-142.

Sklar, A. (1996a). Random variables, distribution functions, and copulas -a personal look backward and forward. In [Rüschendorf, et al. (1996)], pp. 1-14.

Sklar, A. (1996b). Representation of associative functions via simultaneous Schröder equations. *Aequationes Math.* **51**, pp. 157-158.

Smítal, J. (1988). *On Functions and Functional Equations.* Adam-Hilger, Bristol-Philadelphia.

Szabó, G. (1985). Remark on the problem No. 11 of Prof. C. Alsina. *Proc. 23rd Int. Symposium on Functional Equations.* Univ. of Waterloo, Canada, pp. 62-63.

Tardiff, R.M. (1980). On a functional inequality arising in the construction of the product of several metric spaces. *Aequationes Math.* **20**, pp. 51-58.

Tardiff, R.M. (1984). On a generalized Minkowski inequality and its relation to dominates for t-norms. *Aequationes Math.* **27**, pp. 308-316.

Targonski, G. (1981). *Topics in Iteration Theory.* Vandenhoeck & Ruprecht, Göttingen.

Thorp, E. (1960). Best-possible triangle inequalities for statistical metric spaces. *Proc. Amer. Math. Soc.* **11**, pp. 734-740.

Tomás, M.S. (1987). Sobre algunas medias de funciones asociativas. *Stochastica* **XI**, pp. 25-34.

Trillas, E. (1979). Sobre funciones de negación en la teoría de conjuntos difusos. *Stochastica* **III**, pp. 47-59.

Trillas, E. (1980). *Conjuntos Borrosos.* Vicens-Vives, Barcelona.

Trillas, E. (1982). Assaig sobre les relacions d'indistingibilitats. *Actes Primer*

Congrés Català de Lógica Matemàtica, Barcelona, pp. 51-59.

Trillas, E., Alsina, C. and Terricabras, J.M. (1995). *Introducción a la Lógica Borrosa.* Ariel, Barcelona.

Trillas, E. and Valverde, L. (1981). On some functionally expressable implications for fuzzy set theory. *Proc. 3rd Int. Sem. on Fuzzy Set Theory*, Johannes Kepler Univ., Linz, pp. 173-190.

Trillas, E. and Valverde, L. (1982). On indistinguishability and implication. *Proc. 4th. Int. Sem. on Fuzzy Set Theory*, Johannes Kepler Univ., Linz.

Trillas, E. and Valverde, L. (1984). An inquiry into indistinguishability operators. In H.J. Skala et al. (Eds.), *Aspects of Vagueness.* Reidel, Dordrecht, pp. 231-256.

Trillas, E. and Valverde, L. (1985). On implication and indistinguishability in the setting of fuzzy logic. In J. Kacprzyk and R.R. Yager (Eds.), *Fuzzy Sets for Comp. Aided, Manag. Dec. Sup.*, Cologne, pp. 198-212.

Valverde, L. (1984). On the structure of F-indistinguihability operators. Report nr. UCB/CSD 84/200, Univ. California at Berkeley.

Widder, D. (1941). *The Laplace Transform.* Princeton Univ. Press, Princeton, NJ.

Yao, J.Z. and Ling, C.-H. (1964). On the existence of additive generators for a t-norm. *Notices Amer. Math. Soc.* **11**, p. 127.

Zadeh, L.A. (1965). Fuzzy Sets. *Inform. and Cont.* **8**, pp. 338-353.

Zadeh, L.A. (1974). Calculus of fuzzy restrictions. *Fuzzy sets and their applications to cognitive and decision processes* (Proc. U.S.-Japan Sem., Univ. Calif., Berkeley). Acad. Press, New York, 1975, pp. 1-39.

Zadeh, L.A. (1975). Fuzzy logic and approximate reasoning (in memory of Grigore Moisiel). *Synthese* **30**, pp. 407-428.

Zadeh, L.A. (1978). Fuzzy sets as basis for a theory of possibility. *Fuzzy Sets and Systems* **1**, pp. 3-28.

Zadeh, L.A. (1979). A theory of approximate reasoning. In J.E. Hayes, D. Michie and L.I. Mikulich (Eds.), *Machine Intelligence* **9**. Elsevier, Amsterdam, pp. 149-194.

Zimmermann, H.J. (1991). *Fuzzy Sets Theory and Its Applications.* Kluwer, Dordrecht.

Index

Abel's equation, 8, 94, 155, 161
Abel-Schröder system, 167, 169, 170
Aczél's theorem, 2, 82-85
Additive generator, 26, 34, 35, 38-51, 53-57, 72-77, 85-87, 94, 97
Archimedean copula, 44-48
Archimedean t-norm, 16, 23-38, 51
Associative copula, 21, 44-48, 99, 105-106, 110, 177
Associativity equation, 2

Bachelli representation, 87
Bisymmetry equation, 95, 96
Bivariate probability distribution, 18
Boundary conditions of a t-norm, 9
Boundary conditions of an s-norm, 13
Boundary curve of the zero set, 34, 36, 105, 137, 157, 162

C-volume, 197
Cancellative, 82
Cauchy's equation, 7, 41, 160, 190
Classification of t-norms, 66
Complete monotonicity, 199, 200, 205
Composition laws, 80
Concave function, 8
Concave t-norm, 174
Conical t-norm, 137-143, 180
Continuous Archimedean t-norm, 23-38, 66, 70
Continuous non-Archimedean t-norm, 57-64, 66
Convex function, 8, 44, 45
Convex t-norm, 179-181
Copula
  2-dimensional, 4, 17-22, 44-48, 71, 78, 99, 102-106, 153-155, 176, 193, 210
  n-dimensional, 197-203
Craigen-Páles representation, 85

De Morgan triple, 117
De Rham's system, 117
Developable t-norm, 142, 143
Diagonal of a t-norm, 15, 152-155
Differentiable t-norm, 49-51
Directed families, 71
Discontinuous t-norm, 64-70, 80, 89, 90, 211, 212
Distributivity, 134-137
Dominance relation, 182-189
Dual copula, 21, 99
Duality, 110-127, 214

ϵ-close functions, 189
Exchange principle, 123
Extension to a t-norm, 151-171

Falsity Principle, 123
Families of t-norms, 70-81
Flow, 93, 94
Frank's equation, 99-110
Frank's t-norms, 5, 6, 72 #(2.6.5), 99
Frank's theorem, 100-102

Fréchet-Hoeffding bounds, 18, 103, 198
Fuzzy Logic, 4, 5, 117-125
Fuzzy set, 4, 5, 106, 117-125

Generalized associativity equation, 98
Generalized hyperbolic law, 80, 109
Generalized logical connectives, 117-125
Geometrically convex function, 186
Grade of membership, 117-125

Hilbert's 5th problem, 1, 93
Homogeneous associative function, 129-134
Homothetic curves, 141

Idempotent element, 16, 60, 61
Implication function, 123
Independence of the t-norms axioms, 209-210
Inner additive generator, 34, 38-51, 55
Inner multiplicative generator, 39
Iseomorphism, 36
Iterates of functions, 6

Jensen's equation and inequalities, 7, 8
Joint probability distribution, 18, 198

Kimeldorf-Sampson properties, 102-103, 106, 119
Kleene's inequality, 122
Krause representation, 62, 63, 87, 88

Law of the excluded middle, 122
Left-continuous t-norm, 68-70
Level curve, 157
Limiting t-norm, 71, 76-77
Ling's theorem, 3, 88
Lukasiewicz triple, 119

Majorization, 173
Marginal probability distributions, 18, 198
Measure of dependence, 4, 19
Membership function (see Fuzzy set)
Menger's inequality, 4
Metric induced by an associative copula, 126, 127
Midpoint concave t-norm, 175, 178
Midpoint convex t-norm, 179
Minkowski inequality, 182
Minkowski law, 81, 109
Modus Ponens, 124
Multiplicative generator, 38-43

n-box, 196
n-duality, 110-127
n-increasing function, 197
n-interval, 196
n-monotonic function, 199
Nilpotent, 28, 35
Normal triple, 121
Nullnorm, 93

Order among t-norms, 11, 39-42
Ordinal sum, 58-64, 103
Orthomodularity, 118
Outer additive generator, 34, 38-51, 55
Outer multiplicative generator, 39

Pi-Calleja representation, 85-86
Positive t-norm, 203
Positive semidefinite function of order k, 203
Power associativity, 86-89
Probabilistic metric space, 3, 4, 5, 6
Probabilistic relation, 125
Pseudo-inverse, 25, 38, 54

Quasi-arithmetic mean, 96, 97, 98, 134
Quasi-concave t-norm, 174
Quasi-convex t-norm, 179
Quasi-copula, 22
Quasi-homogeneous t-norm, 130-131
Quasi-inverse, 90

Radial symmetry, 104
Rational Archimedean copula, 150

Rational Archimedean t-norm, 143-150
Representation theorems for t-norms, 23-38, 57-64, 81-89, 212-213
Right-inverse, 25
Ruled t-norms, 137-143

s-norm, 6, 13, 37-38
Schröder's equation, 8, 87, 87, 152, 161, 170, 196
Schur-concave t-norm, 173, 176
Schur-convex t-norm, 178, 180
Sections of a t-norm, 17, 155-171
Serial iterates, 194-196, 217
Set of uniqueness, 158-171, 215
Simultaneous associativity, 5-6, 99-110, 213
Sklar representation, 89
Sklar's theorem, 18, 197-198
Smooth convex t-norm, 181
Strict involution, 111
Strict s-norm, 13
Strict t-norm, 11, 34, 38
Strong negation, 111
Stronger binary operation, 11
Subadditive function, 8, 39-41
Summand of ordinal sum, 58
Superadditive function, 8
Superadditive t-norm, 178, 216
Survival copula, 104

t-conorm, 6
T-indistinguishability operator, 125
t-norm, 4, 5, 6, 9-17, 23-51
T-powers, 13
T-transitivity, 125
Theory of information without probability, 3
Totally positive function of order k, 203
Translation equation, 93
Triangle function, 4, 106
Triangle inequality, 4, 5, 6
Triangular norm (see t-norm)

Ulam-Hyers inequality, 190-193
Uniformly close associative functions, 181-194
Uninorm, 84

Weak dominance relation, 188, 204, 205
Weaker binary operation, 11
Weighted mean, 96, 97, 98
Weighted quasi-arithmetic mean, 96-97, 134
Wiener-Shannon law, 81,

Zero set, 34, 36, 137, 163